T0297236

Statistical Principles and Techniques in Scientific and Social Investigations

Statistical Principles and Techniques in Scientific and Social Investigations

Wojtek J. Krzanowski
is Emeritus Professor
of Statistics at the
University of Exeter, UK.

OXFORD
UNIVERSITY PRESS

OXFORD

UNIVERSITY PRESS

Great Clarendon Street, Oxford OX2 6DP
United Kingdom

Oxford University Press is a department of the University of Oxford.
It furthers the University's objective of excellence in research, scholarship,
and education by publishing worldwide. Oxford is a registered trade mark of
Oxford University Press in the UK and in certain other countries

First published 2007
Reprinted 2013

Published in the United States of America by Oxford University Press
198 Madison Avenue, New York, NY 10016, United States of America

British Library Cataloguing in Publication Data
Data available

Library of Congress Cataloging in Publication Data
Data available

ISBN 978-0-19-921309-2

To Wendy, Adam, Helen and Mike

Contents

Preface

Everyday life at the start of the twenty-first century is very different from what it was a hundred, or even fifty, years ago. Enormous changes and developments have taken place in virtually every sphere of human activity: work, leisure, travel, health, commerce, finance, the list is almost endless. Many of these changes have been prompted by scientific development and innovation, but equally many by conclusions drawn from careful observation of social, commercial or industrial processes, and yet others by painstaking and sometimes lengthy experimentation. In all of these areas—science, experimentation and observation—statistical reasoning and statistical methodology plays a central role. This is because our world is governed by chance, and a concomitant of chance is variability. There is variability between individuals in virtually any activity or study that we care to focus on, but nevertheless there are recognisable overall patterns when we aggregate individuals. However, the result is that we can draw very few conclusions with certainty; they generally have to be qualified in some way. Statistical reasoning has been developed over the years in order to quantify the inherent uncertainty in our conclusions, and statistical techniques have been devised in order to enable us to extract as efficiently as possible the underlying patterns from among the prevalent variability.

A statistician is someone who has been trained in the art and science of statistical reasoning, and who is thus able to assist workers in a great variety of subject areas to draw the best conclusions they can from their studies. This is not, perhaps, the first image that comes to mind when one hears the word 'statistician'. The popular stereotype as conveyed by the media would probably be one of two alternatives: either someone who spends his or her time collecting endless facts about events in particular sports (e.g. numbers of wickets taken and balls bowled in cricket; numbers of shots on target in football; numbers of home runs in baseball) or a government official compiling large tables of figures (such as social security payments; unemployment rates; household expenditures on different commodities). These individuals certainly do exist, and such tasks are indeed carried out, but they form only a very small part of the world of statistics. Moreover, it is a part that will not concern us in this book. The collection of data, and their systematic organisation into tables or charts, is of course important. Undoubtedly it is sometimes badly done, and a book focusing specifically on the correct methods would of course be useful—but not perhaps either very contentious or exciting. What is much more challenging and sometimes controversial is the elucidation of *evidence* from collections of facts, the extraction of pattern from the variation and the quantification of the uncertainty attaching to conclusions; in short, the real stuff of statistics.

But why have a relatively non-technical guide to statistical reasoning and techniques? Would it not be sufficient to let the trained statistician get on with his or her job, and to accept the conclusions when they are presented in a final report? The simple answer is of course that this would be unsatisfactory from the perspective of the researcher who wants to be involved at all stages of the work. However, the art and science of statistics has developed to such an extent over the past hundred years that it now comprises a very complex collection of techniques and methodology, and it is quite difficult for the 'interested amateur' to build up a reasonable working knowledge in a short space of time. Quantitative data now permeate most subjects so that many individuals have to acquire at least a rudimentary knowledge of statistics, but many elementary texts and courses in statistics focus overmuch on detailed mathematics and computations. Not only do many students find mathematics a barrier, but also such an approach often fails to provide an overall structure and hence a genuine understanding of the subject. By stepping back and taking an overview, the necessity for detail may become more apparent and the detail itself may become easier to absorb. On the other hand, perhaps the potential student does not want detailed knowledge of methods to apply to data, but would rather prefer to gain an appreciation of the reasoning processes and the rationale for the methods in order to make good sense of analyses that have been reported in the media or in the more technical literature. It is hoped that this volume can address such needs.

However, I must admit to having two other potential types of reader in mind, with more of a missionary intent. One is the diehard sceptic: the scientist or researcher who has never had any time for statistical analysis, either believing his or her research to be so clear cut as not to need any such mumbo-jumbo or not really understanding the basis of the analysis and so disbelieving the results. I have encountered this attitude occasionally during a lifetime as a statistician, perhaps less so latterly but certainly once quite recently while listening to an interview on early morning radio. A study had been conducted into various 'alternative' medicines, and statistical analysis had shown that there was no evidence to suggest that these were more efficacious *in general* than traditional medicines. The interviewee, a worker at the clinic where the study was conducted, dismissed the results on the grounds that 'the statistical calculations were too complicated to understand, and had been done by someone who had never even donned a white coat or seen the effects on patients of some of these medicines'. This worker seemed to miss the point on several counts. If he found statistics too difficult to understand, on what basis was he rejecting the results of the analyses? If the statistician had worn a white coat while doing the computations in the clinic, would that have made the results acceptable? But, most pertinently, is not the *overall* pattern of results with these medicines, and their comparison with traditional ones, of more relevance than close first-hand knowledge of a few individual cases? Would not it therefore be worth investing a little effort into trying to understand the statistical calculations? Perhaps, an account of general statistical reasoning as well as the rationale for specific techniques in non-technical language may help to overcome the understanding barrier, and convince some of these sceptics that statistics does have something to offer.

Even though the incidence of the diehard sceptic is much less now than it was when I first came into the profession nearly 40 years ago, by contrast the self-appointed expert now appears to be far more prominent than was once the case. The fault here, perhaps, is the provision of many short courses in statistics, often geared to researchers in some specialised substantive area, together with easy access to many statistical software packages that will carry out complicated analyses at the press of a button. While the researchers may feel that they have studied the subject in some depth, often they have not progressed beyond some fairly basic ideas and techniques. Unfortunately, the old adage that a little learning can be dangerous is as true today as it was originally, and while the computer software will crunch the numbers efficiently enough it will not be able to say whether or not the buttons pressed were the correct ones for the problem in hand. The end product can all too frequently be an inappropriate analysis, which is then vigorously defended and the results accepted—especially if the 'statistician' is indeed expert but in some other field. This can lead to the sort of disastrous consequences that have recently had exposure in the media, for example in connection with the 'cot-death' cases in the British courts. It is my hope that a relatively non-technical account of statistical reasoning will help to build a more solid foundation; even if it does not temper any self-delusions about extent of knowledge of the subject then maybe it will at least bolster that knowledge and reduce the risk of incorrect analyses.

So these are my excuses for putting pen to paper. As should be evident from the foregoing, my objective is to set out the basic ideas underlying most of the major areas of statistics and to survey the most popular statistical techniques, but without becoming bogged down in computational or mathematical technicalities. My hope is that readers will thereby acquire some insight into what statistics is all about, will understand when particular techniques should or should not be used, and will perhaps gain the confidence to go on to the more technical aspects in other books. I have to admit to finding statistics an endlessly fascinating subject, not least because it has so many diverse component parts: philosophy and logic as well as mathematics and computing; art as well as science; and applicability to so many different areas of endeavour as well as theoretical rigour. If the book conveys something of this fascination then it will have served its purpose.

For the first five chapters the emphasis is mainly on the reasoning behind the techniques and much less on the techniques themselves, and then this emphasis is reversed in the remaining five chapters. Of course, it is impossible to exclude mathematics entirely from such a project, so some formulae, mathematical notation and mathematical ideas are inevitably present. However, I have tried to keep them at a reasonably rudimentary algebraic level, and have not strayed into any areas of advanced mathematics (such as differential or integral calculus, in particular). In order to give a linking thread to the development, I have organised the chapters in a loosely historical fashion—although I make no claims for this to be in any way a genuinely historical development. Readers looking for that will have to consult much more specialised texts.

In conclusion, I would like to thank Steve Fisher, Jim Kay and Don Webster for providing many helpful comments and suggestions on a first draft of the book, also Alison Jones, Carol Bestley and Jessica Churchman at OUP for seeing it through the publication process.

Wojtek J. Krzanowski
Exeter, UK

1 Probability

Introduction

Much, if not most, statistical reasoning is built upon a foundation of probability; so if a good understanding of statistical principles is to be gained, then the basic ideas of probability must first be understood. This is the reason why many introductory books on statistics begin with a survey of probability theory. However, an equally compelling reason might be the increasing importance that probability in its own right is playing in the world about us, and the fact that some facility with probability calculations or assessments can be invaluable in our day-to-day lives. Most people are perfectly comfortable with simple probability statements, such as that there is an even chance of obtaining a head as opposed to a tail on the spin of a coin, but once familiar physical frameworks are abandoned then the ground quickly becomes shakier. Even a seemingly straightforward assertion in a weather forecast that there is a 20% chance of rain in a particular region can seem ambiguous, while intuition leads many people completely astray when the situations become more complex or conditions are imposed on the events in question.

So let us start this chapter by considering some everyday situations that require assessment of probabilities, and suggesting the intuitive responses that they might elicit. Then during the course of the chapter, as we cover the relevant principles that enable the correct values to be calculated, we can demonstrate how intuition can fail when dealing with probability and hence highlight the importance of acquiring a working appreciation of the subject. These situations encompass a spectrum of types, ranging from the purely recreational to the very serious in which a correct assessment of probabilities is vital.

At the purely recreational end of the spectrum, a host or hostess organising a cocktail party might wonder how many people need to be invited if there is to be a better than even chance that at least two people present will have a shared birthday. A reader who has not come across this 'birthday problem' before would probably not be able to make a numerical guess but would intuitively expect the answer to be 'very many', so perhaps the actual number might turn out to be a surprise. At the same end of the spectrum, but now a problem that was faced in real life by contestants in an American TV game show, is what has become known as the 'Monte Hall problem'. Monte Hall was the host of the show, and the game was very simple. The contestant was faced with three doors concealing various prizes. Behind one of the doors was a valuable prize (e.g. a brand new top-of-the-range limousine) while behind the other two doors there was something considerably less attractive (e.g. a can of beans). The contestant was invited to choose one of the doors, but before the prize was revealed Monte Hall would open one of the other two doors. Knowing which door concealed the valuable prize, he always opened a door that revealed the unattractive prize. The contestant was then given the opportunity to switch doors if he or she wished, so the question to the reader is: Would you switch if you were the contestant? Many people would argue that there was an equal chance that each door concealed the valuable prize when the initial choice was made, and there was now also an equal chance that each of the remaining doors concealed it. So there would be nothing to be gained by switching. We shall see whether this reasoning is correct or not.

Moving on to situations that require more precise assessment of probabilities, diagnostic tests are now quite commonplace in medicine, engineering and other areas, so here are two possible scenarios. In the first, your neighbour comes to see you and he is very worried. He has had a medical check-up, in the course of which he has been screened for a rare disease. About one person in a thousand suffers from this disease, and the equipment used for the diagnosis is known to be 99% accurate. He has tested positive for the disease, but the treatment for it is extremely expensive so he wants to know your assessment of the chance that he actually has the disease. The second scenario, in an engineering context, concerns an aircraft in flight: the warning light comes on in the cockpit if the landing gear is faulty, and from past experience it is known that it only fails to come on about once in a thousand times when the landing gear actually is faulty. On the other hand, it is also known that it comes on about five times in every thousand flights when the landing gear is not faulty, and the landing gear is likely to be faulty on about four flights in every thousand. During a particular flight the warning light comes on, so the captain has to assess the chance that the landing gear is faulty in order to make a decision about where to land. What would you advise? Of course, both of these situations involve various factors that lie outside the realm of pure probability and that might influence the final decision (such as cost of treatment, seriousness of disease, safety of passengers, etc.), but even if these factors are disregarded the issues are now more complicated. However, intuition would probably suggest that there is a high chance that your neighbour has the disease, and that the landing gear is faulty.

The final situation is the most critical one, and relates to the legal case alluded to in the preface in which a mother was jailed for life, having been convicted of killing her two children. Her defence was that the deaths were both natural cot deaths, but at the trial an expert paediatric witness gave evidence that the chance of a mother's experiencing two natural cot deaths was one in 73 million. This seemed to be such a small value that the jury rapidly reached a guilty verdict. It was subsequently established that this probability had been badly miscalculated and should more correctly be quoted as one in just over 130,000, but even this seems to be a very small chance indeed. Would you have gone along with the jury in deciding that such an event was so unlikely that she must have been guilty? If you would, then you also would have been a victim of what is known as the *prosecutor's fallacy*—in fact, the conviction was subsequently overturned at the second appeal and after much extra detail from experts in probability. Although not every legal case involving probability arguments is as dramatic as this one, nevertheless increasing numbers of cases rest on probability assessments and members of juries now have to be prepared to make judgements based on them.

We will return to each of these situations in more detail at appropriate points during this first chapter.

Foundations of probability theory

In a relatively informal sense, the notion of chance, games of chance, and chance mechanisms for aiding decision-making have all been around for thousands of years. Brian Everitt gives an entertaining brief history in the first chapter of his book *Chance Rules* (1999), and cites various milestones along the way: Palamedes inventing games of chance during the Trojan wars to bolster the soldiers' morale, references in the *Bible* to the drawing or casting of lots to ensure a fair division of property or privileges, and documented instances of the throwing of dice or similar for purposes of divination. Whereas the latter reasons for the use of chance mechanisms have fluctuated in and out of fashion depending on judicial systems and religious proclivities, games of chance have had a solid and unbroken line of practice since the time of the Trojan wars. The original chance mechanism of tossing animal bones has been gradually refined over many years, and we are now very familiar with such gambling accessories as dice, cards, roulette wheels, lottery balls and electronic random number generators.

Constant practice at some specific activity inevitably gives the practitioner an intuitive 'feel' for, and increasing skill in, that activity. The inveterate darts player, for example, can do in a split second the mental arithmetic required to sum three numbers, often first having to double or treble a few of them, and then subtract them from a three-digit number on the scoreboard—whereas the same person might have considerably greater trouble in adding up the week's grocery bill in the supermarket and determining the correct change from a hundred pounds, because the context of the arithmetic has changed.

So also with games of chance. Over the years, players would garner experience and develop intuition about the chances of particular outcomes in particular circumstances, and then use this intuition to hone their skill and improve their prospects of monetary gain. No attempt seems to have been made for many years, however, to systematise this intuition in any way or to develop any form of mathematical theory; surprisingly many years, indeed, considering the great progress made in so many other areas of mathematics during this time.

The big advance came in the seventeenth century, with the first mention (reputedly by Galileo and other Italian mathematicians) of the concept of 'equally likely possibilities'. It seems to be something of a paradox that the breakthrough in the definition and subsequent development of the calculus of probability is itself based on something that implicitly involves this concept ('equally likely'). However, on closer scrutiny, this is not really much of a drawback. It applies specifically to the sorts of situation prevalent in games of chance, where there is but a limited set of possibilities associated with any particular operation, and the symmetry or nature of this operation dictates that no single possibility has preference over any of the other possibilities. Thus it seems to be intuitively perfectly reasonable that, when a six-sided die is thrown, any one of the six faces is as likely to appear as any of the others; when a 'fair' coin (i.e. one which is not weighted in any way) is spun, then it is as likely to come down 'heads' as 'tails'; when a standard deck of 52 cards has been thoroughly shuffled, then any 1 card is as likely to be pulled out from the deck as any of the 51 others, and so on.

All this seems so obvious that one is led to wonder why it was not formally recognised for so long, particularly as so much probability theory flows directly from it. Be that as it may, it is a fundamental part of probability calculations so it forms our starting point below. Although it is evident from our introductory section above that probability calculations can arise in many different spheres of life and contexts, the fundamental rules and procedures were developed with games of chance specifically in mind. Hence we will use this framework to establish and illustrate the concepts, but will turn to the earlier examples for everyday applications.

Sample space; definition of probability of an event

The first, and crucial, step is to list (traditionally within curly brackets) the set of all equally likely possibilities for any operation; this is usually termed the *sample space* for that operation. Some examples of such sample spaces are as follows.

1. If the operation consists of throwing a die and noting the number of spots on its upper face, then the sample space is {1, 2, 3, 4, 5, 6}.

2. If the operation is the spinning of a coin and noting its upper face, then the sample space is {head (H), tail (T)}.
3. If either two coins are spun together, or one coin is spun twice in succession and both upper faces are noted, then the sample space is {HH, TH, HT, TT}.
4. Finally, if a standard pack of cards is shuffled, one card is drawn from it and its suit and value are both noted, then the sample space is {2♣, 3♣, ..., 10♣, J♣, Q♣, K♣, A♣, 2♦, ..., A♦, 2♥, ..., A♥, 2♠, ..., A♠}, where ♣ (Clubs), ♦ (Diamonds), ♥ (Hearts), ♠ (Spades) are the suits, 1, 2, ..., 10 are the numerical values; and J(ack), Q(ueen), K(ing), A(ce) are the pictures.

We are then generally interested in calculating probabilities of specific *events*, and any event comprises one or more of the individual possibilities making up the sample space. Some examples of events for each of the sample spaces above are as follows:

1. 'Throwing an even value': possibilities 2, 4 and 6.
2. 'Obtaining a tail': possibility T.
3. 'Obtaining one head': possibilities HT and TH.
4. 'Drawing a picture card': possibilities J♣, Q♣, K♣, A♣, J♦, Q♦, K♦, A♦, J♥, Q♥, K♥, A♥, J♠, Q♠, K♠ and A♠.

The probability of any event is then just the number of possibilities making up this event, divided by the total number of possibilities in the sample space. Once the notion of equally likely possibilities has been accepted, this definition of probability is intuitively self-evident. The probabilities of the above events are therefore:

$$\frac{3}{6} = \frac{1}{2}; \quad \frac{1}{2}; \quad \frac{2}{4} = \frac{1}{2}; \quad \frac{16}{52} = \frac{4}{13}.$$

While this seems to be very straightforward, it is important to note that correct calculation rests upon appropriately listing the equally likely possibilities in the sample space, as well as correctly identifying which of them corresponds to any given event. The first of these is often the trickier one to get right, and this is where many people go wrong when calculating probabilities. As a simple example, consider the probability of getting one head with the spin of two coins. An incorrect approach would be to say that the sample space consisted of the set of possible values of 'number of heads', viz. {0, 1, 2}, so the event 'one head' comprises just one of the 3 possibilities and hence has probability $\frac{1}{3}$. But the flaw here is that the possibilities in this 'sample space' are *not* equally likely, as the value 1 can occur in two ways so is twice as likely as either 0 or 2. The only correct sample space here is the one given in (3) above, so the probability is $\frac{1}{2}$.

A couple of consequences flow directly from this definition. First, if the event in question corresponds to none of the possibilities in a sample space of n items (e.g. drawing a 'joker' from the pack of cards) then its probability is $\frac{0}{n} = 0$. Second, if the event comprises all the possibilities in the sample space (e.g. obtaining either a head or a tail with the spin of a coin) then its probability is $\frac{n}{n} = 1$. The values 0 and 1 correspond to impossibility and certainty respectively, and they constitute the lower and upper limits of the probability scale with any probability value lying between them. (This is the scale used in probability calculations, rather than the percentage scale 0–100% with which some readers may be more familiar.) If an event were detrimental to us then we would hope its probability was very close to 0, while if it were a very beneficial event we would hope its probability was close to 1. We will come back to this scale of values for probability again, a little later in this chapter.

Essentially, the above calculation of probability holds good for any event associated with equally likely possibilities, however complicated the events and sample spaces may be. Games of chance, and other chance mechanisms for decision-making, usually ensure that sample spaces of equally likely possibilities can be readily specified. The problems arise when these sample spaces, and their associated events, become very extensive or difficult to identify.

The Monte Hall problem, in fact, can be solved using this approach and is one of the more straightforward instances. Here the sample space just consists of the three possible arrangements of prizes behind doors. Let us denote the attractive prize by a and the unattractive prize by u. Then the three rows below denote the three possible arrangements, so constitute the sample space:

Door 1	Door 2	Door 3
a	u	u
u	a	u
u	u	a

Whichever door the contestant has chosen, in two out of the three possibilities (the two in which the chosen door conceals the unattractive prize) Monte Hall has no further choice when picking a door concealing the unattractive prize, so therefore leaves the attractive prize behind the remaining (unchosen) door. In these two possibilities, the contestant who switches is bound to gain the attractive prize. In the other possibility the contestant's choice conceals the attractive prize, so Monte Hall can open either of the other doors to reveal the unattractive prize and the other door also conceals the unattractive prize. So in this case, the contestant who switches will lose the attractive prize. Now let E_1 be the event 'winning if the contestant does not switch' and E_2 be the event 'winning if the contestant switches'. From the argument above we see that the number of possibilities in E_1 is just 1 out of the three in the sample space so its probability is $\frac{1}{3}$, but the number of possibilities in E_2 is 2 so its probability is $\frac{2}{3}$. By switching, the contestant actually doubles his or her original chance of winning the attractive prize!

More complicated examples involving very large sample spaces are readily provided by the national lottery, and this is in fact a source of almost endless questions about probability. The player is invited to choose six numbers between 1 and 49 (no number can be repeated, and the order in which they are chosen is unimportant), and wins the jackpot if the six chosen numbers exactly match the six thrown up by the chance mechanism used by the game's organisers. Suppose that a player is considering picking a run of six numbers in sequence (e.g. 24, 25, 26, 27, 28, 29). A pertinent question then might be: What is the probability that the winning set of numbers is a run of six consecutive ones? Here the chance mechanism is such that any set of six numbers between 1 and 49 is as likely to occur as any other, so the sample space S consists of all sets of six such numbers, while the event E we are interested in is 'a run of six consecutive numbers'. To obtain the required probability, we thus need to find out how many sets of six numbers constitute each of S and E. It is fairly straightforward to list the ways in which event E can occur, namely the sequences $(1, 2, 3, 4, 5, 6)$, $(2, 3, 4, 5, 6, 7)$, $(3, 4, 5, 6, 7, 8)$, ..., $(44, 45, 46, 47, 48, 49)$. So there are 44 possibilities in E, but how many equally likely possibilities are there in S? Writing them all out and counting them would be beyond human capabilities, and in such cases we must resort to appropriate mathematical techniques for a solution. It is not our intention to go into mathematical details in this book, so the interested reader should consult the topics of *permutations* and *combinations* in a specialist text to find results concerning the selection, arrangement or counting of objects. In the present instance, application of the appropriate rules from the theory of combinations establishes that there are 13,983,816 equally likely sets of six numbers in S. So the probability of E is $44 \div 13,983,816$, or approximately 1 in 320,000. Of course once the player has made a particular selection, then that player's chance of winning is just the probability of *that particular* set of numbers being selected, or approximately 1 in 14 million—the same as any other player's chance!

Conditional probability, independence and the multiplication rule

We have seen that some convenient mathematical theory can help us when we have to determine the number of possibilities in a sample space, and by extension the number of these possibilities making up an event, in complex situations or large sample spaces. An alternative approach in these, and similar, situations is to break up the events of interest into a series of sequential or connected events. To illustrate the idea, consider drawing two cards at random from a standard deck of 52.

However, first let us pause briefly to note the phrase 'at random', which we have used here for the first time but which is used very frequently in statistics and will recur often throughout this book. It is shorthand for 'such that all possibilities are equally likely', interpreted in a broad sense. Thus by choosing two cards at random we assume that the pack has been shuffled thoroughly and there is no reason to suppose that any pair of cards is more likely to have been chosen than any other pair. If I were to say that I had chosen a sample of citizens of a town 'at random', it would generally mean that I had used some form of chance mechanism on a list of citizens (e.g. an electoral roll) in such a way that any individual in the list has the same chance of being chosen as any other. We will return to the topic of random sampling later; for the present we simply use the phrase 'at random' to avoid having to use the much clumsier longhand all the time.

So let us return to the random selection of the two cards. We might now ask the question: What is the probability that both cards are spades? We can use the methods already described to determine the value of this probability as follows. First, the sample space consists of all the possible ways in which we can draw two cards from the pack of 52. Any of the 52 cards can be picked first, and with each of these choices any of the 51 remaining cards can be picked second to give $52 \times 51 = 2652$ possibilities. But these are possibilities listing first choice and second choice distinctly, whereas when we look at the two chosen cards it does not matter which was chosen first and which second. Since every $(1, 2)$ choice will have a $(2, 1)$ counterpart, the number of distinct pairs is $2652 \div 2 = 1326$. By similar reasoning, the number of these possibilities that consist of two spades is $13 \times 12 \div 2 = 78$. So the probability that both chosen cards are spades is $\frac{78}{1326} = \frac{1}{17} = 0.0588$. In other words, we would expect to get two spades when drawing two cards at random from a pack about once in 17 such draws.

However, we can obtain this probability perhaps more easily by considering the two draws separately and specifying what conditions need to be satisfied for the event of interest to happen. In order to obtain two spades with the two draws, we must obtain a spade first and then *another* spade with the second draw. If we now want to calculate the probabilities of the two events, the second draw must acknowledge that one card has already been drawn, and that that draw has resulted in a spade. So if we denote by A the event 'the first card drawn is a spade' and by B the event 'the second card drawn is a spade', we need to compute first the probability of A, and then the probability of B given that A has occurred. This latter is a *conditional* probability, that is, a probability that acknowledges that a specified condition has been satisfied. If we write $P(A)$ for the probability of A, then traditionally $P(B|A)$ is written for the conditional probability of B given that A has occurred.

These two probabilities can be found very easily from the basic principles. For the first draw, there are 52 cards equally likely to be selected, 13 of which are spades. So $P(A) = \frac{13}{52} = \frac{1}{4}$. Then, *once A has occurred*, there are 51 equally likely cards left, 12 of which are spades. So $P(B \mid A) = \frac{12}{51}$. Note that this probability is not the same as $P(A)$, because the sample space and the number of possibilities that constitute B have both changed as a consequence of A happening. Nevertheless, the two probabilities are both computed very easily. Moreover, we note that if we multiply the two together we obtain $\frac{1}{4} \times \frac{12}{51} = \frac{3}{51} = \frac{1}{17}$, the same as the probability we previously obtained for the event that 'the first two cards drawn are spades', that is, for the joint occurrence of A and B. This is no coincidence, as it is in fact an example of the multiplication rule of probability: For *any* two events A and B, $P(A \ \& \ B) = P(A)P(B \mid A)$ [$= P(B)P(A \mid B)$, since it is immaterial which way round we label the events].

The multiplication rule opens up the route to much easier calculations of probabilities than laborious listing or calculation of sample spaces and 'favourable' possibilities. This is because by breaking a complex event down into a sequence of simple interlinked ones we can calculate the probability of each simple event fairly readily, and then use the multiplication rule to combine these separate probabilities. Moreover, the multiplication rule generalises to more than two events in the obvious way, by conditioning at each step on all the events contained in previous steps. Thus with three events A, B, and C: $P(A \ \& \ B \ \& \ C) = P(A)P(B \mid A)P(C \mid A \ \& \ B)$; with four events: $P(A \ \& \ B \ \& \ C \ \& \ D) = P(A)P(B \mid A)P(C \mid A \ \& \ B)P(D \mid A \ \& \ B \ \& \ C)$, and so on. For example, if we want the probability that the three cards chosen at random from a deck of 52 are all spades, we simply need to multiply the previous probability that two chosen at random were spades by the conditional probability that the third is a spade *given that the first two were spades*. In the latter case, there are 50 equally likely cards left, 11 of which are spades, so P(third card is a spade | first two were spades) $= \frac{11}{50}$ and P(all three are spades) $= \frac{1}{17} \times \frac{11}{50} = \frac{11}{850} = 0.0129$.

While the multiplication rule can simplify calculations considerably, the concept of conditional probability can sometimes cause problems, as we shall see later. However, in some situations there is further simplification. Suppose that instead of drawing two cards from the same pack, we had drawn one card from each of two packs and again want to find the probability that we had drawn two spades. In this case, the probability of the first card being a spade is again $\frac{13}{52} = \frac{1}{4}$, but now the probability of the second one being a spade *given that the first one was a spade* is also $\frac{13}{52} = \frac{1}{4}$. This is because there are still 52 cards and 13 spades in the second pack, irrespective of what card was drawn from the first pack. So now the probability of drawing two spades is $\frac{1}{4} \times \frac{1}{4} = \frac{1}{16}$, and not $\frac{1}{17}$. The difference comes because the occurrence of the first event has no effect on the probability of the second event happening. In other words, in this case we have

$P(B \mid A) = P(B)$. In such situations we say that the events A and B are *independent*, by contrast with the *dependent* events we previously had, and the multiplication rule simplifies considerably because now $P(A \,\&\, B) = P(A)P(B)$. Note that the use of two different packs of cards in this example is the same as using just one pack but replacing each card back in the pack and shuffling before drawing the next card. Previously we did not replace the card before drawing again. In general, sampling *with replacement* induces independent events, while sampling *without replacement* leads to dependent events. Such operations as sequentially tossing a coin or throwing a die also clearly lead to independent events, as the outcome of one throw will not influence the outcomes of any others. Thus the probability of getting, say, two sixes in two throws of a die is $\frac{1}{6} \times \frac{1}{6} = \frac{1}{36}$, which is, of course, also easily seen by listing the 36 possible equally likely outcomes (remembering that a throw of 1 first and 2 second is distinct from 2 first and 1 second), and then identifying (6, 6) as the only one of the 36 outcomes that corresponds to the event in question.

The idea of independent events and independence is thus a great simplifying factor in probability calculations, but it is also a considerable trap for the unwary. The naive approach to probability is to assume that all events are independent and to use the simple rule $P(A \,\&\, B) = P(A)P(B)$ even when it is not correct to do so. Sometimes this is of no consequence (the difference between a chance of 1 in 16 and 1 in 17 is perhaps negligible to the human eye in the context of the example above), but sometimes it can have serious repercussions. This was so in the case referred to earlier, of the eminent paediatrician who naively assumed independence of events when computing the probability of two cot deaths in a family. His reasoning that this probability was simply the product of the probability of one cot death with itself led to the figure of 1 in 73 million, which contributed substantially to the false imprisonment of the mother. So one of the most important things to do when carrying out probability calculations is first to examine the nature of the events in question and determine carefully whether they are dependent or independent.

The addition rule and mutually exclusive events

When two events A and B are under consideration, the probability of their joint occurrence is only one possible focus of interest. Very often we might want to find the probability that *either* one *or* the other (or, indeed, both) occur. When there are more than two events, this generalises to finding the probability that *at least* one of them occurs.

To establish the general idea, consider a simple example. Suppose we draw a single card from a standard deck, and want to know the probability that it is either a spade or an ace (or both). We can find this by writing down all the 52 possibilities, and then noting that 16 of them are either spades or aces. How would we arrive at this number without writing down all 52 possibilities? Well, we could add the number of spades (13) to the number of aces (4) to obtain 17. However, we would then note that one of the 13 spades was the ace, and one of the 4 aces was a spade. So we have counted the ace of spades twice, and we must subtract one of these two instances from 17 to get the right answer of 16. The probability of either an ace or a spade is thus $\frac{16}{52} = \frac{4}{13}$. This reasoning holds good no matter how many of the possibilities in the sample space 'overlap' the two events A and B. When we add the number of possibilities nA making up A to the number of possibilities nB making up B, we include the number in the overlap, nAB say, twice. So we must remove one of these sets of nAB in order to obtain the correct number in the event 'either A or B (or both)', which is therefore $nA + nB - nAB$. But dividing these numbers by the number n in the sample space converts each one to a probability of the corresponding event: $P(A) = nA/n$, $P(B) = nB/n$, and $P(A \ \& \ B) = nAB/n$, and this provides the addition law of probability: $P(\text{either } A \text{ or } B) = P(A) + P(B) - P(A \ \& \ B)$. [From now on, we use the 'inclusive or' whereby 'either A or B' automatically implies 'or both'.]

Just as independent events simplify the multiplication rule, so the addition rule can be simplified in certain cases. If the events A and B can never occur together, then nAB must be zero and hence $P(A \ \& \ B)$ must be zero. Such events are said to be *mutually exclusive*, and in this case the addition rule has the simple form $P(\text{either } A \text{ or } B) = P(A) + P(B)$. For example, if we want the probability that a single card drawn from a deck of 52 is either a heart or a spade, then this is just $\frac{13}{52} + \frac{13}{52} = \frac{26}{52} = \frac{1}{2}$, as a single card cannot be both a heart and a spade.

One immediate consequence flows from this simplified version of the addition rule, regarding the *complement* A^c of an event A. A^c is usually read 'not A', and consists of all the possibilities that are not in A. For example, if A is the event 'an odd number is obtained with a throw of a die', then A^c is the event 'an even number is obtained'; if A is the event 'a spade is drawn from a standard deck' then A^c is the event 'either a club, diamond, or heart is drawn', and so on. Clearly, A and A^c are mutually exclusive, so $P(\text{either } A \text{ or } A^c) = P(A) + P(A^c)$, and one or the other *must* happen so $P(\text{either } A \text{ or } A^c) = 1$. Equating the right-hand sides of these equations thus establishes that $P(A^c) = 1 - P(A)$.

The results established thus far enable us now to solve the 'birthday' problem outlined in the introduction. This asks us to find the probability that *at least* two people in a room have the same birthday. Whenever the phrase 'at least' occurs, it is usually easiest to first find the probability of the complementary event. So here we first ask the question: What is the probability that all the people have *different* birthdays? To answer this, we first need to make a couple of simplifying assumptions, namely that there are 365 days in each year (thus ignoring problems introduced by leap years) and that any chosen individual's birthday is equally likely to be one of these 365 days (thus ignoring any 'seasonality' effects in birthday patterns). We return briefly to these assumptions below, but first let us do the calculations.

Consider each individual in the room in turn, in any order, and focus on their birthdays. The first chosen person can have any birthday, but for every subsequent individual to have a different birthday from each of the previously chosen people, the number of 'permissible' days on which that birthday can fall decreases by one. By the equally likely assumption, the probability that the new person's birthday differs from all those of the previous people is just the number of 'permissible days' divided by 365. So the probability that the second person chosen has a different birthday from the first one is $\frac{364}{365}$, the probability that the third person chosen has a birthday different from the first two is $\frac{363}{365}$ and so on. So, by the multiplication rule of probability, if there are n people in the room then the probability that they all have different birthdays is

$$p_{\text{diff}} = \frac{365}{365} \times \frac{364}{365} \times \frac{363}{365} \times \cdots \times \frac{365 - n + 1}{365}.$$

Thus, since the event 'at least two people have the same birthday' is complementary to the event that they all have different birthdays, the probability we require is $p = 1 - p_{\text{diff}}$. We can evaluate p on a computer (or pocket calculator) for successive values of n, finding, for example, 0.0027, 0.0082, 0.0164, 0.0271 for $n = 2, 3, 4$ and 5. Continuing with this process, the first value greater than 0.5 is 0.5073 when $n = 23$. So we just need 23 people in a room to have a better than even chance that there will be (at least) one shared birthday among them.

This may seem to some readers to be a surprisingly small number. Of course, it is only a 'better than even chance' rather than certainty, and even with 40 people in the room there is still a probability of 0.1088 that all birthdays are different. Moreover, we have made some simplifying assumptions, so the reader might wonder if this is a critical consideration. In fact, ignoring leap years has made negligible difference (as using 366 in place of 365 above gives $p = 0.5063$ for $n = 23$, and only one year in four has 366 days), while seasonality of birthdays is likely to *decrease* the required number of people (as it acts in the direction of increasing the likelihood of coincidence). So this is one example that illustrates the way in which intuition can be misleading.

To return to generalities, extending the addition rule to more than two events is very straightforward if all the events are mutually exclusive, as then $P(A \text{ or } B \text{ or } C) = P(A) + P(B) + P(C)$ and so on. However, if the events are not mutually exclusive then the rule becomes more complicated, because the probabilities of all possible pairs, triples and so on have to be accommodated. For example, with three events A, B, C the rule is:

$$P(A \text{ or } B \text{ or } C) = P(A) + P(B) + P(C) - P(A \& B) - P(A \& C)$$
$$- P(B \& C) + P(A \& B \& C).$$

while for more than three events the pattern established here is continued, namely add probabilities of all single events, subtract probabilities of all joint pairs, add probabilities of all triples, subtract probabilities of all quartets, and so on. However, this approach to calculating probabilities of complex events is rather intricate. A more direct approach is provided by the next result.

The total probability theorem

We have seen above that the two events A and A^c are mutually exclusive, and one or other is certain to occur. In general, any sample space can be divided up into a set of events that are mutually exclusive, and that between them cover all the possibilities in the sample space. Such events are said to be *exhaustive*. For example, if one card is drawn from a standard deck, then the four events: 'the card is a club', 'the card is a diamond', 'the card is a heart', and 'the card is a spade' are exhaustive because each event comprises 13 mutually exclusive possibilities in the sample space, and between them the four events use up all the 52 possibilities in this space.

Let us therefore suppose that, in a given situation, the k events $B_1, B_2, B_3, \ldots, B_k$ constitute such an exhaustive division of the sample space, and A is *any* event we are interested in. Then the total probability theorem simply states that

$$P(A) = P(A \& B_1) + P(A \& B_2) + P(A \& B_3) + \cdots + P(A \& B_k),$$

the sum being over all the exhaustive events. Applying the multiplication rule to the joint probabilities on the right-hand side of this equation gives the operationally more useful form:

$$P(A) = P(A \mid B_1)P(B_1) + P(A \mid B_2)P(B_2) + P(A \mid B_3)P(B_3)$$
$$+ \cdots + P(A \mid B_k)P(B_k).$$

The very useful feature of this theorem is that it can be applied to any event A, and the partition of the sample space can be chosen to suit the problem under consideration. To illustrate the utility of the theorem, let us consider a relatively complicated but popular game involving dice. (The following calculation is certainly not an easy or 'obvious' one for the novice, but demonstrates how an apparently very difficult problem can be solved quite readily if a systematic approach is adopted.) In the game of craps, the gambler plays against the casino. The gambler rolls a pair of dice, and the sum of the two numbers is the important quantity. If the sum is either 7 or 11 the gambler wins immediately, but if the sum is 2, 3 or 12 then the casino wins immediately. If the sum is 4, 5, 6, 8, 9 or 10, then that number is called the gambler's 'point', and the gambler keeps rolling the dice until either a 7 or the 'point' appears. If a 7 appears before the 'point' then the casino wins, but if the 'point' appears before a 7 then the gambler wins. What is the probability that the gambler wins?

Clearly, the event A of interest here is 'the gambler wins', but what is the sample space and what would be the most convenient partition into events $B_1, B_2, \ldots$? Since the total score on the first roll of the dice determines the outcome of the game, the set of equally likely outcomes of rolling two dice should constitute the sample space. There are 36 such outcomes, given by the 36 combinations $(1, 1), (1, 2), \ldots, (1, 6), (2, 1), \ldots, (2, 6), \ldots, (6, 6)$ of paired values for each die. But since it is the sum of the two values that is important, then the most convenient partition of the sample space would be into the 11 sums $2, 3, 4, \ldots, 11, 12$. They are clearly mutually exclusive and exhaustive; let us therefore denote the sum of the scores on the first roll of the dice by B. We can find the probabilities of each of these scores by counting up how many of the 36 possibilities in the sample space gives the requisite score. Thus $B = 2$ results from just the one possibility $(1, 1)$, $B = 3$ from either $(1, 2)$ or $(2, 1)$, $B = 4$ from $(2, 2)$, $(1, 3)$ or $(3, 1)$, and so on. Dividing the number of possibilities by 36 gives the probability of each value of B, and these are found to be as follows:

B	2	3	4	5	6	7	8	9	10	11	12
$P(B)$	$\frac{1}{36}$	$\frac{2}{36}$	$\frac{3}{36}$	$\frac{4}{36}$	$\frac{5}{36}$	$\frac{6}{36}$	$\frac{5}{36}$	$\frac{4}{36}$	$\frac{3}{36}$	$\frac{2}{36}$	$\frac{1}{36}$

So these values provide the values of $P(B = i)$ in the total probability theorem. We now need the conditional probabilities $P(A \mid B = i)$, that is, the conditional probabilities that the gambler wins given that the first roll gave a score i. Since the gambler wins outright when the first roll is either 7 or 11, $P(A \mid B = 7)$ and $P(A \mid B = 11)$ both equal to 1. Conversely, the casino wins outright if the first roll gives 2, 3 or 12, so $P(A \mid B = 2)$, $P(A \mid B = 3)$ and $P(A \mid B = 12)$ must all be zero. What about each of the other values of B? Consider $B = 4$. The gambler keeps rolling until the score rolled is *either* 7 (when the casino wins) *or* 4 (when the gambler wins). But there are six ways of rolling a 7 and three ways of rolling a 4. So there are nine outcomes that terminate the game, three of which yield a win for the gambler. Hence $P(A \mid B = 4) = \frac{3}{9}$. Similar reasoning shows that when $B = 5$ there are 10 ways of terminating the game, 4 of which yield a win for the gambler, and when $B = 6$ there are 11 ways of terminating the game, 5 of which yield a win for the gambler. Symmetry shows that $B = 8$ is the same as $B = 6$, $B = 9$ is the same as $B = 5$ and $B = 10$ is the same as $B = 4$. So the set of values of $P(A \mid B = i)$ is as follows:

B	2	3	4	5	6	7	8	9	10	11	12
$P(A\mid B)$	0	0	$\frac{3}{9}$	$\frac{4}{10}$	$\frac{5}{11}$	1	$\frac{5}{11}$	$\frac{4}{10}$	$\frac{3}{9}$	1	0

Thus, finally, applying the total probability theorem we multiply each $P(A \mid B)$ in this table by the corresponding $P(B)$ from the previous table and add up the results to get the probability that the gambler wins, that is,

$$P(A) = 0 + 0 + \frac{3}{9} \times \frac{3}{36} + \frac{4}{10} \times \frac{4}{36} + \frac{5}{11} \times \frac{5}{36} + 1 \times \frac{6}{36}$$

$$+ \frac{5}{11} \times \frac{5}{36} + \frac{4}{10} \times \frac{4}{36} + \frac{3}{9} \times \frac{3}{36} + 1 \times \frac{2}{36} + 0,$$

which turns out to be 0.4929. This is a result that is sufficiently close to 0.5 to give the impression that the gambler has an 'even chance' of winning any single game, but sufficiently far away from 0.5 to ensure that over the long run of games that the casino plays against different gamblers, it will win sufficiently more than it loses in order to ensure that it makes a healthy profit!

Bayes' theorem

Reverend Thomas Bayes holds a central place in the theory of statistics, despite the fact that by the time of his death in 1761 he had not published any technical or philosophical notes or results. His fame rests on a single posthumous paper and the consequent theorem bearing his name, which as a deductive result in probability theory is wholly uncontroversial but when extended later to inferential statistics has been responsible for some of the most heated arguments of the twentieth century. We shall look at the inferential aspects in due course, but for the present let us confine ourselves to the deductive result in probability theory. In its simplest guise it is just a consequence of the multiplication rule being expressible in two ways: if A and B are any two events, then the joint probability $P(A \,\&\, B)$ can be written *either* as $P(B \mid A)P(A)$ *or* as $P(A \mid B)P(B)$. Setting these two expressions equal to each other, therefore, and dividing both sides of the equation by $P(A)$, we obtain the result:

$$P(B \mid A) = \frac{P(A \mid B)P(B)}{P(A)}.$$

This is just a straightforward way of deriving one conditional probability from its converse, and in the context of the sort of gambling scenarios we have been considering, it is easily applicable. As an example, consider the game of craps again. The calculations above have used A as the event 'the gambler wins', and the various values of B as the events of getting specified totals on the first two rolls of the dice. The probabilities of each B outcome are given in the first table above, the probabilities of the gambler winning conditional on each B outcome are given in the second table above, and we have calculated that the probability the gambler wins, $P(A)$, is 0.4929. Suppose we enter the casino at the point at which the gambler throws his hands in the air and shouts 'hurrah' upon winning. We might then speculate about what scores had been observed on the first throw of the dice. For example, we might ask what is the probability that a 4 was thrown first. Since we know already that the gambler has won, we are really inquiring about the value of $P(B = 4 \mid A)$. By Bayes' theorem above, this is just $P(A \mid B = 4) \times P(B = 4) \div P(A)$. Substituting the appropriate values from the two tables above, we find this probability to be $\frac{3}{9} \times \frac{3}{36} \div 0.4929$, or 0.056. Using corresponding calculations on each of the other outcomes on the first throw, we obtain the following probabilities:

B	2	3	4	5	6	7	8	9	10	11	12
$P(B \mid A)$	0	0	0.056	0.09	0.128	0.338	0.128	0.09	0.056	0.114	0

Thus, having observed the gambler winning, we can say that the most likely score on the first throw was a 7, with 6, 8 and 11 fairly close to each other (but some way behind 7) as the next contenders.

It is perhaps worth noting at this point that care must be taken if calculations are undertaken from a starting point of conditional probabilities, as this is an area in which mistakes or incorrect interpretations are often made (see, for example, the *prosecutor's fallacy* mentioned in the introduction and discussed below). In particular, $1 - P(A \mid B)$ is equal to $P(A^c \mid B)$ and *not* to $P(A \mid B^c)$, which is a common mistake. So, from the table above, $1 - 0.338 = 0.662$ is the probability of the first throw being anything except a 7 given that the gambler has won, and *not* the probability of the first throw being a 7 given that the casino has won (which of course must be zero, as the *gambler* wins if the first throw is a 7). Confusion over which event is of interest and which is the fixed condition is probably one of the chief sources of error in such calculations.

Moreover, although Bayes' theorem in the above guise falls strictly within the realms of probability theory, notice a subtlety in the interpretation of the two conditional probabilities that gives a hint of the potential controversies lying ahead. The various $P(A \mid B)$ values are probabilities of (future) winning, given (present) values of the first throw, but the various $P(B \mid A)$ values are probabilities of (previous) first throw scores given (present) winning. So the former relate to probabilities of what *might happen in the future*, whereas the latter relate to probabilities of what *might have happened in the past* and thus led on to the present state of affairs. So, despite an ostensibly deductive probability calculation, we are in fact using the latter result as a guide to *inference*. We shall pick up this point again later on.

For the present, we can now complete the calculations of the remaining three problems posed in the introduction. The first two (medical diagnosis and landing gear failure) are straightforward applications of Bayes' theorem so can be dealt with fairly quickly, while the third one focusing on the cot deaths bears some extra discussion.

First the medical diagnosis. Let $+$ denote a positive and $-$ denote a negative result with the screening equipment, and let D denote 'disease' while H denotes 'disease-free'. We are told that about one person in a thousand suffers from the disease, so from the outset we have $P(D) = 0.001$ and $P(H) = 0.999$. We are also told that the equipment is 99% accurate, and we can assume that this means with respect to either possible result, so that $P(+ \mid D) = P(- \mid H) = 0.99$. Consequently, $P(- \mid D) = P(+ \mid H) = 0.01$ are the probabilities of incorrect diagnosis. Your neighbour has had a positive result and wants to know your assessment of the chance he has the disease. Thus you need to evaluate $P(D \mid +)$, and by Bayes' theorem this is $P(+ \mid D)P(D) \div [P(+ \mid D)P(D) + P(+ \mid H)P(H)]$. Inserting the relevant values we find $P(D \mid +) = 0.99 \times 0.001 \div [0.99 \times 0.001 + 0.01 \times 0.999] = 0.00099 \div 0.01098 = 0.0902$. Thus the neighbour's chance of having the disease is still less than 1 in 10.

Likewise for the landing gear problem, let $+$ denote appearance of the warning light and $-$ denote its non-appearance, while F denotes that the landing gear is faulty and W denotes that it is working. The information we are given is that the warning light only fails to come on when it should about once in a thousand times, so $P(- \mid F) = 0.001$ and hence $P(+ \mid F) = 0.999$, but it comes on about five times in a thousand when the landing gear is not faulty, so that $P(+ \mid W) = 0.005$ and hence $P(- \mid W) = 0.995$. Moreover, the landing gear is likely to be faulty on about four flights in a thousand, so $P(F) = 0.004$ and hence $P(W) = 0.996$. When the captain sees the warning light, the assessment required is of $P(F \mid +)$, and by Bayes' theorem this is $P(+ \mid F)P(F) \div [P(+ \mid F)P(F) + P(+ \mid W)P(W)]$. Inserting the given values we have $P(F \mid +) = 0.999 \times 0.004 \div [0.999 \times 0.004 + 0.005 \times 0.996] = 0.003996 \div 0.008976 = 0.445$. Thus there is nearly an even chance that the landing gear is faulty.

Thus, in the above two cases the actual probabilities are lower than most people would intuitively guess, and for the medical diagnosis problem very much lower. This is because most people intuitively think that $P(+ \mid D)$ and $P(+ \mid F)$ are the two relevant probabilities, rather than $P(D \mid +)$ and $P(F \mid +)$. This is the essence of the *prosecutor's fallacy*, the term used for such incorrect reversal of events in a conditional probability and so called because it was first formally identified in legal contexts. The computation of the correct conditional probabilities requires consideration of all the other factors, such as $P(+\mid H)$ and $P(+\mid W)$, and Bayes' theorem makes due allowance for these factors.

So now let us consider the third situation, namely the legal case involving the two cot deaths. This case has provoked written contributions from a number of mathematicians and statisticians, and we draw here on some results described in the articles 'Beyond reasonable doubt' by H. Joyce (2002; http://plus.maths.org/issue21/features/clark/) and 'The power of Bayes' by D.F. Percy (2005). Let us denote the evidence (two deaths) by E, and the guilt or innocence of murder on the part of the mother by G and I respectively. Let us also accept that the chance of two cot deaths occurring to the same mother, that is, $P(E\mid I)$, is 1 in 130,300 as argued by Joyce. The prosecutor's fallacy leads to interpretation of this probability as the probability of innocence given two deaths, that is, as $P(I\mid E)$, and hence suggests that the probability of guilt given the two deaths is 130,299 divided by 130,300—extremely close to 1.

Of course, the appropriate procedure is to use Bayes' theorem to obtain the correct probability, namely $P(I\mid E)$. However, the present calculation is more awkward than the two earlier examples because assessment of some of the relevant probabilities is not so clear-cut. We have already seen that assessment of $P(E\mid I)$ is open to debate and controversy, and for the purpose of this illustration we have taken Joyce's estimate of $\frac{1}{130,300}$. Also, if the mother is guilty then the two deaths will clearly occur so $P(E\mid G)=1$, but what about the probabilities $P(G)$, $P(I)$ of guilt or innocence? Joyce refers to Home Office statistics on numbers of children being murdered by their mothers each year, and by making some conservative estimates of the chances of a mother committing two such murders comes up with the estimates $P(G)=1/216,667$ and $P(I)=216,666/216,667$. Then application of Bayes' theorem yields

$$P(G\mid E)=\frac{1\times\dfrac{1}{216,667}}{1\times\dfrac{1}{216,667}+\dfrac{1}{130,000}+\dfrac{216,666}{216,667}}=0.38.$$

Thus, once we use reasonable estimates of the various probabilities, and apply the correct procedure for combining them via Bayes' theorem, it appears that the mother is about twice as likely to be innocent as she is to be guilty. Moreover, the conservative nature of some of the probability assessments, and the inclusion of material evidence presented at the trial, would argue for even greater probability of innocence. Fortunately these types of arguments prevailed at the second appeal, and the original conviction was overturned.

Relative frequency, subjective probability, and an axiomatic structure

The events we have considered so far, and the probabilities calculated about them, have mostly assumed the existence of a sample space of equally likely possibilities. This is fine for situations such as games of chance and gambling, and was thus perfectly adequate for the classical set-up of the seventeenth century when the subject was first studied. However, as interest widened, it became evident that this classical approach could not cope with more than a fairly limited set of situations, so other approaches to probability would be necessary.

An intuitively reasonable approach is provided by the *relative frequency* approach: the probability of an event is the proportion of times that this event has occurred on all occasions exhibiting 'similar' sets of circumstances to the prevailing ones. The problem, of course, rests in the definition of 'similar'. For example, learner drivers facing a driving test might wonder about their chances of passing it so might wish to calculate the proportion of tests that had previously ended with the driver passing. The simplest procedure would be just to find the total number of tests taken to date in that locality, and find how many of them ended in success for the driver. But there might be different success rates under different prevailing conditions, so a potential examinee might want to tailor the estimate more particularly to himself or herself. But what conditions or circumstances need to be included? Presumably the particular examiner is important, and probably the test circuit. Maybe also the day of the week, as road conditions might vary from day to day. But how about weather? Or the time of day? Or any number of other such factors? The problem is that inclusion of more and more conditions reduces the number of previous occasions that match them, thereby reducing the number of test occasions for the calculation. Estimating a probability from small numbers of occasions leads to volatility: one instance more or less than the number observed can produce big swings in the proportions. So the relative frequency idea is fine, provided one is calculating the proportion from a large enough base, and provided that one accepts some leeway in the definition of 'similar occasions'. However, it is reassuring to notice that the equally likely definition will accord with the relative frequency approach: If we have the patience to draw a card from a well-shuffled deck thousands, or millions, of times, we will find that each card is drawn in a proportion of them that is very close to $\frac{1}{52}$. Indeed it was the relative frequency approach, using Home Office statistics as indicated above, that enabled reasonable assessments of the probabilities $P(G)$ and $P(I)$ to be reached in the cot deaths case.

Nevertheless, there remain many events in everyday life to which we may want to assign a probability value, but which neither fall within the remit of equally likely possibilities nor have a large sequence of previous occurrences to which we can apply the relative frequency argument. A large proportion of such events are personal, governing our everyday actions. What is the probability my train will be late? What is the probability that I will be able to buy the compact disc I want in this shop? What is the probability that I will be able to get two tickets for the concert tonight? We are always assessing our chances in regard to such matters, and most of us have developed sufficient intuition based on knowledge and experience to be able to estimate such chances at least roughly. If we went further and were prepared to assign an actual value between 0 and 1 to a specified event, then that value would be our *subjective probability* of the event in question. Of course, there is no reason to suppose that someone else's subjective probability of the same event would be the same as our value; indeed, if a room full of people were to be asked to assign a probability value to a particular event then there would probably be nearly as many different suggested values as people in the room. However, if each individual is consistent in the way that he or she assigns such subjective probability values, then there is no reason why we should not apply all the usual rules of probability to these values. We shall see later that a whole body of techniques of statistical inference accommodates the possibility of employing subjective probabilities.

The crucial point is that individuals must be consistent in the way that they assign probability values, so how do we ensure this? The matter was settled by the twentieth century Russian mathematician Kolmogorov, who in 1933 published the *axioms* of probability. Any probability model is guaranteed to give consistent results from probability calculations as long as it obeys these axioms. There are surprisingly few of these axioms, they are very simple ones, and all remaining results of probability theory (such as those described throughout this chapter) are derivable from them. In straightforward language, the axioms require

1. any assigned probability value to lie between 0 and 1;
2. the value 1 to represent certainty of an event and
3. the assigned probability of at least one out of a set of events to equal the sum of the assigned probabilities of each event, providing these events are all mutually exclusive.

This last condition must hold however many events there are in the set, whether this number is finite or infinite, and is clearly the most testing one to satisfy in the case of subjective assignment of probabilities. But the probabilities in the equally likely and the relative frequency approaches automatically satisfy the axioms.

The above does not claim to be either an exhaustive account nor indeed anywhere near a comprehensive coverage of probability theory. Nevertheless, some technicalities have been considered in a little bit of detail. This is because probability seems to be a fundamental concept that most people are prepared to think about and dabble in, but one that presents many potential pitfalls and one where wrong answers to problems are regrettably more common than right answers. So it is hoped that the basic ideas of probability have been conveyed in to a sufficient extent for the reader to be aware of the principles behind the requisite calculations. Having established these principles we have a firmer basis from which to explain statistical reasoning, so we can now go on to develop some statistical ideas.

Populations, Samples and Data Summary

Introduction

We live in an age when data gathering is commonplace. Most readers will at one time or another have been asked to fill in a questionnaire, or to assist in an investigation, or to take part in a survey, or to collect data on some topic of interest. Recent examples to have come the author's way have included a questionnaire from the local theatre to elicit information on frequency of attendance, preferences in programming and attitude towards ticket prices; a doorstep survey on awareness of various commercial products included in advertising campaigns; some research into the relative popularity of different degree programmes involving mathematics at UK universities; and an unexpected invitation by the local supermarket to participate in a tasting session designed to evaluate customers' preferences in soft drinks. In all such cases, the objective is to collect *data* of some form on a *target* set of individuals, and then to *summarise* the information that has been gathered in the hope that it will guide future action. The theatre will want to put on the sort of programmes at the sort of prices that will maximise their attendances; the supermarket will want to tailor its soft drink holdings in order to maximise profits and so on. In this chapter, we introduce some of the fundamental terminology and concepts that underlie such data gathering exercises, and thereby provide a basis for the statistical ideas to come.

Populations

At the heart of any statistical investigation is the idea of a *population* of individuals. This is simply the target set of all possible individuals about whom information is sought. These individuals can be either animate or inanimate, and the strictness with which they are defined should mirror the aims of the investigation or data gathering exercise. Thus the UK census conducted every 10 years targets the population of all individuals living in the United Kingdom at the time, and attempts to gather a vast range of information from each individual. By contrast, the theatre survey mentioned above would aim to target the population of all potential attendees of the theatre, and seeks only information on a fairly limited range of topics. In a study of the effectiveness of a new drug in lowering blood pressure of subjects suffering from hypertension, the target population would be all individuals who suffer from hypertension and the information gathered would focus on blood pressure measurement before, during, and after taking the drug. Naturally, if the focus were gender-related, then the target population would be either all males suffering from hypertension or all females suffering from it. As an example of an inanimate population, a surveyor may be conducting an investigation of small commercial properties (say ones whose total floor area is less than some fixed value) within a given distance of a city centre. In this case, the target population would be all commercial properties of the requisite size within the requisite distance of the city centre.

Identifying the target population is one critical aspect of any study, but associated with the population are the *observations* or *measurements* that are to be made on each individual of that population. These comprise the *data* for study. There are several different *types* of measurements, and since the different types generally demand different statistical techniques for their analysis, we need to consider them briefly here. The traditional division is into four different types, distinguished according to the amount of numerical information that they contain.

At the lowest level we have observations that contain no numerical information, but simply record either the presence or absence of a particular attribute of the individual, or the category into which we can place that individual. For example, we might note whether individuals in a study wear glasses or not; we may categorise an individual according to the colour of their eyes (blue, brown, green, grey); or we may describe a commercial property as single-storey, two-storey or other. Such an observation is said to be at the *nominal* level of measurement. Note that a numerical value can be associated with each category for the purposes perhaps of *coding* the categories, if the data are to be entered into a computer say, but such numbers are not amenable to the usual arithmetic operations. For example, the above eye colours could be coded 1 for blue, 2 for brown, 3 for grey and 4 for green, so that a database on a computer could store the eye colours of many people very efficiently; but it would be plainly nonsense to calculate the average of these numbers and to announce that there was an 'average' eye colour of 2.74 in the population!

Some observations produce categories that have an implied order or that can be ranked according to some underlying graduation. For example, a market research survey might ask respondents to indicate how often they purchase a particular product by ticking one of the categories 'never', 'occasionally' or 'frequently'; a doctor might record the amount of pain suffered by a patient on the scale 'mild', 'moderate' or 'severe'; and the director of a theatre might ask questionnaire respondents to indicate which of the following age bands they fall in: 0−18; 19−30; 31−60 or greater than 60. Such observations are said to be on an *ordinal* scale; not only are the individuals classified into categories, but they are also *ranked* in terms of the characteristic in question. Note, however, that there is no implication that there are equal intervals between the ranks: we cannot make any assumptions about *how* much more pain there is in the 'severe' category than in the 'moderate' category, or *how* much more often the 'frequently' purchased article is than the 'occasionally' purchased one. Thus, if numerical codes are associated with each category, it is still not valid to apply arithmetical operations to them. In some cases where the categories have an actual numerical basis, as with the age in the theatre survey, then choice of appropriate numerical codes might allow arithmetic summaries, but in general this should be treated with caution.

Once an observation involves either counting or recording a numerical value of some kind on each individual, then the usual arithmetic operations *are* appropriate. For example, we might record an individual's height in metres, weight in kilograms and the number of trips he or she has made by bus in the previous month; or we might count the number of rooms in each commercial property included in the survey and record the proportion of them used for storage purposes. Counts are of course restricted to be integer values, while measurements such as heights or weights are only limited by the accuracy of the measuring instrument and can in principle take any value within some range. Measurements that can only take a finite set of distinct values are said to be *discrete*, while those that can take any value within a specified range are *continuous*. Arithmetic operations are perfectly acceptable on either type, so we can talk about an average height, weight, number of trips made by bus in a month, or number of rooms. Of course, the average of a discrete measurement will not usually itself be within the same discrete set of values (e.g. an average of 2.78 bus trips per month), but that is no drawback to interpretation. Numerical measurements have a unit of measurement that has equal intervals between successive values, but not necessarily a true zero point that indicates absence of the quality being measured. Equal intervals means that the difference between values of, say, 50 and 60 represents the same amount of the quantity being measured as that between 80 and 90, while a true zero point means that a value of 40, say, is twice as much as 20. Temperature in Fahrenheit has equal degree intervals but no true zero point (as 0°F does not imply absence of temperature), so the difference between 60° and 90° is the same as between 60° and 30°, but 60° is not twice as hot as 30°. Any such measurements are said to be on an *interval* scale. However, if a measurement additionally has a true zero then it is said to be on a *ratio* scale. Examples include familiar measurements such as length, height or weight: a distance of 2 km is twice as far as

1 km, and a sack weighing 30 kg is three times as heavy as one weighing 10 kg.

While the distinction between the different scales of measurement is both interesting in its own right and important when developing statistical methodology, we will not be concerned with such a fine gradation. For practical purposes it is usually sufficient to be able to distinguish between *categorical observations* (i.e. the nominal + ordinal scales) and *numerical measurements* (i.e. the interval + ratio scales) in order to determine which statistical techniques are admissible in each particular application, so we will restrict ourselves to this coarser distinction. In addition, we may also need to distinguish between discrete and continuous measurements from time to time.

So let us return to the idea of a population. If we were able to gather the observations or measurements of interest on all individuals in a population, we would notice variation in values from individual to individual. In fact, for any specific type of observation or measurement, there would be a *distribution* of values. For example, heights and weights would vary considerably from individual to individual, but some values (e.g. around 1.8 m for height and 70 kg for weight) would be much more prevalent than other values (e.g. around 2.1 m and 90 kg). Likewise, some eye colours are more likely than others; some people travel much more often on buses than others, and so on. The main objectives of any statistical study centre on determination of important features of these distributions. Thus the theatre questionnaire was aimed at determining which types of performance (drama, comedy, thriller, musical, etc.) are the most popular among all likely patrons, while a survey of commercial properties might be conducted in order to determine the range of floor areas of such properties, or the most common number of rooms in them.

The basic problem with all the above, of course, is that for the collected information to be of true worth we need to determine the feature of interest for the *whole* population. Unfortunately, the population in question is usually either too big or too difficult to pin down exactly to make the measurement of each individual feasible. For example, what exactly *is* the whole population of potential patrons of a particular theatre? Can we ever collect the whole population of sufferers from hypertension? Even if we can identify and potentially contact or obtain each member of the population, the process of doing so and then obtaining the requisite measurements is usually prohibitively expensive so this is not a practical proposition. Occasionally, the perceived importance of the study overrides expense, as in the national census conducted every 10 years in the United Kingdom, when a great deal of expense is incurred in trying to contact every individual in the whole population and extract a large amount of information from each one. Collection of full and accurate information is deemed essential to the national interest and to the conduct of government, but this is an exceptional case. Mostly, it is only possible to measure a small portion, or *sample*, of the population, and then to hope to extract information about the totality from the measurements obtained from this small portion. The mechanics of so doing constitute the subject matter of *statistical inference*, viz. inferring features of the population from a sample. In order to ensure that this process is not futile, we must ensure that the sample we choose is somehow 'representative of the whole population'. Although the practice of sampling and arguing from samples goes back several hundred years, it was only during the last century that systematic study was undertaken into methods of ensuring such representativeness of a population. Since it is so important in drawing correct conclusions, we now briefly survey the various available methods.

Samples and sampling

Given the need in most studies to sample from the target population, the two most pertinent questions are: How large should the sample be? How should the sample be obtained? The first of these questions needs some appreciation of sampling variability before it can be answered, so we will come back to it later; for the present we will focus on the second question.

Several intuitively appealing procedures suggest themselves, and are indeed used in practice, but all include pitfalls that must be fully appreciated. The first one is to go for administrative convenience or pragmatism, and simply to choose the sample that is easiest to obtain. This is known as *accessibility sampling*, and can be effected by asking for volunteers, or selecting the individuals closest to hand, or using some convenient mechanism that requires little effort to implement. The great danger here is that such a method, unless it is very carefully controlled and administered, will not produce a sample that is truly representative of the population but rather one that is skewed towards some particular portion of the population. Such a skew is generally referred to as a *bias*, and hence such a sample is called a *biased* sample. Asking for volunteers is perhaps an obvious way of generating a biased sample; the more outspoken members of a population, or the ones holding extreme views, are more likely to volunteer than are members of the 'silent majority', so the tendency is for the outcome to represent the more extreme views. A slightly more hidden generator of bias is the use of some convenient but perhaps inappropriate mechanism. An example of this occurred in the early days of pre-election opinion polls, when a poll of 50,000 people before the 1948 US presidential election produced the forecast that Truman (a Democrat) would receive 44% of the vote, Dewey (his Republican opponent) would receive 50% and the independents Wallace and Thurmond between them the other 6%, thus giving Dewey a comfortable victory. The actual outcome was almost exactly the reverse, with 50% for Truman, 45% for Dewey and 5% for the two independents. What went wrong? The answer is that the poll was conducted by telephone, and in those early days this was still quite an expensive appliance. As a consequence, Republican supporters were more likely to have telephones than Democrat supporters, and being skewed towards the Republicans in this way biased the sample and produced a useless prediction.

A second approach, recognising the inherent dangers of accessibility sampling, is to ask the researcher drawing the sample to exercise his or her subjective choice in producing a 'representative' sample. This is often termed *judgemental sampling* or *purposive sampling*, and it is hoped that the researcher will by this means eradicate any anticipated biases. This it will almost certainly do, but unfortunately it will not counteract either unanticipated biases or ones introduced by the researcher's personal prejudices or lack of knowledge of important aspects of the population and its relationship to the measurements being made. Barnett gives a good example of the latter in his book *Elements of Sampling Theory* (1974), in a study of ash content in coal lying in different piles near the mine. The researcher, being aware of possible differences in the coal contained in different piles, samples the coal by taking some from the edge of each pile. This ensures a representative sample of the different piles—but the ash content varies with the size of the lumps, and the ones on the edges of the piles are likely to be smaller lumps. So the researcher will not have a 'representative' sample in terms of ash content!

Variations on the theme have therefore been devised, either to cope with particular difficulties or to try and eliminate biases. For example, when attempting to conduct a survey of a population that is difficult to access (e.g. the homeless) a *snowball sample* might be contemplated: each person taking part is asked to recruit some acquaintances, who then recruit a few of their acquaintances, and so on. Of course a possible source of bias here is towards the views of those participants who have the most acquaintances, but on the other hand this may be the only way to reach the desired population. Another long-established possibility is the *quota* sample. This is a favoured approach for surveys conducted in the streets of busy towns and cities (often on Saturdays when shopping tends to be at its height). The researcher who designs the study with a given target population in mind considers the structure of that population carefully, draws up a set of categories within the population that it is important to represent, and gives each interviewer a quota of the different categories to interview (e.g. x young women with children, y old-age pensioners, z middle-aged men, and so on). It is then up to the interviewer to accost the most likely looking passers-by in order to fill the quotas. Here the most likely sources of bias will come from personal preferences or prejudices on the part of the interviewer (selecting those people who they think will be the most sympathetic, perhaps).

However, all of the above methods are subjective and their main drawback is the fact that there is no objective way of determining how representative of the target population the resulting samples have turned out to be. This is a major worry if the results of the survey prove unpalatable to the researcher, because it is then only too easy to dismiss the results on the grounds that the sample might be biased. In order to avoid this charge we must have some mechanism that ensures representativeness of the sample, and in order to do this we must appeal to probability and to the idea of randomness.

Random samples and some variants

The only sure way of eliminating the possibility of bias, whether this bias is anticipated or not, is to remove the selection of the sample from the hands of the researcher and to make it purely objective. Moreover, no individual or group of individuals should have any preference over any other in this process. This goal can be achieved by using some chance mechanism for the selection of the sample members, and imposing the requirement that every individual in the population has an equal chance of selection. As an intuitive idea, this goes back a long way. Examples can be found as early as the sixteenth century in which rudimentary attempts at random selection are evident, and Laplace in his 1802 study of the number of births in France describes a procedure approximating to the idea of equal chance of selection for various sections of the populace. However, the formal structure was not established until the twentieth century. The great statistician Ronald Aylmer Fisher propounded the concept of randomisation in connection with the design of experiments, and we shall examine its use in this context in a later chapter. Its subsequent implementation in sampling and surveys opened the way to the information-gathering explosion of the second half of the century. The implementation is very straightforward: If we want a random sample of k individuals we just list the target population, and then by using a chance mechanism that allocates equal probability to each member of the list, select the requisite number k of members. If the members of the population are numbered from 1 to n, the simplest chance mechanism is a bag containing n balls numbered 1 to n that is shaken thoroughly and from which k balls are drawn. The numbers on the balls indicate the list members to be included in the sample. This is a very familiar device, used frequently in choosing prize winners in raffles or postal competitions, as well as in making up the schedule of sporting knockout competitions such as the annual Association Football (FA) Cup. More sophisticated chance mechanisms have been introduced over the years, in order to cope with large populations more easily. A popular device now is one based around *pseudo-random numbers*, which are numbers generated from mathematical series in such a way that any k-digit number obtained in this way has the same chance of occurring as any other k-digit number. Such numbers were originally listed in large tables, to be read off by the researcher conducting the survey, but now feature in electronic software on the most basic computers so can be invoked at the touch of a button. Readers who have ever purchased premium saving bonds will be familiar with this idea, because the monthly winning numbers are drawn by just such a pseudo-random number generator (nicknamed on its inception as ERNIE). An example of the use of pseudo-random numbers in producing a random sample would occur if the Council of a large town wished to conduct a survey of the inhabitants in order to gauge their views on provision of leisure facilities in the town. The electoral roll provides an easily accessible list of all the voters in the town; an alternative list might be provided by a listing of all streets and houses by number within the streets. Whichever list is taken, the items on it are first numbered from 1 to n. If a sample of size k is desired, then k pseudo-random numbers in the range 1 to n are generated and the people or houses that coincide with these numbers on the list are surveyed.

However, while such a simple random sample will avoid conscious or unconscious biases, it will only be truly 'representative' of a homogeneous population and it may badly skew the results in the case of a highly structured population. For example, consider a survey that is aimed at estimating the average monthly wage bill of a large company. Suppose there are 1000 employees, and a sample of 100 workers is to be taken. It is easy to list the population and to take a random sample as described above. But suppose that of the 1000 employees, 50 are in top executive positions, 150 are middle managers and the remaining 800 are on basic administrative and clerical grades. Here we have three distinct types (or 'strata') of worker, and the typical monthly wage will be very different for each stratum. A simple random sample of 100 workers, while correctly implemented, may end up with 20 top executives, 40 middle managers and 40 others, thus badly over-estimating the monthly wage bill. For the sample to be truly representative, it should contain the same proportion of each stratum of worker as there is in the whole population, so there should be 5 top executives, 15 middle managers and 80 others. The correct procedure here would therefore be to take random samples of these sizes from *within* each stratum. This is known as a *stratified random sample*. Whenever there are such identifiable strata in a population, this procedure should be adopted. Of course, the proportions in each stratum may not always be known beforehand, in which case either a guess can be made or the sample can be split evenly across the strata. While this may not give the most accurate result, at least it should be a considerable improvement over a simple random sample.

Another situation where a simple random sample is not ideal, but perhaps this time for economic reasons, is when the population is much dispersed spatially. Consider an investigation into running costs in National Health Service hospitals in a particular Health Authority. Typically the Authority will be split into different regions, each region will have a number of different hospitals, and each hospital will have a possibly large number of wards, with probable different costs for different wards. The most relevant target population is the set of all possible wards. This population can certainly be listed, and a simple random sample taken as outlined above, but such a sample might well contain just one ward from each of a large number of different hospitals spread right across the whole authority. If the study involves collecting detailed information about different sorts of costs, it will probably necessitate travelling to each individual sampling point to elicit this information and this will involve a lot of time and expense. A better scheme, and one which provides no reduction in accuracy, would be to randomly choose a given number of regions, then for each chosen region randomly choose a given number of hospitals, and then for each chosen hospital randomly sample a given number of wards. The final selection of wards would thus be much more localised, and this scheme would involve far less travel and expense than the simple random sample. This is known as a *cluster sample*, and can be used whenever the population has a *hierarchical* structure of the above type.

There are, of course, many other possible variations or adjustments that can be made in particular or specialised circumstances, but the above probably encompass the great majority of sampling schemes adopted in practice. The vital ingredient is the idea of random sampling, as this eliminates all subjectivity and associated biases, and this concept will turn up again very centrally when we move on to consider statistical inference. Meanwhile, in the remainder of this chapter, we follow up the methods of sampling by looking briefly at the ways in which we can handle the outcomes of the sampling process. Having designed a sampling process to elicit information in some area of interest, we are inevitably faced with the question of what to do with the resulting data! This is of course an extremely wide topic, and it is not the aim of this book to cover detailed methods. However, some of the fundamental quantities are going to be essential for understanding later discussion, so we give a brief description of them here.

Categorical measurements: cross-tabulations

Since categorical data do not lend themselves readily to arithmetic operations, the most efficient summary is a cross-tabulation display of the incidences in each of the various possible combinations of categories. Such a cross-tabulation is often called a *contingency table*, and it enables direct comparison to be made of either totals or proportions in different categories. To illustrate the general idea, we give the cross-tabulation produced by a study into three-year survival of 754 breast cancer patients. The observations taken on each patient were: age (under 50, 50−69, over 70); diagnostic centre (Tokyo, Boston, Cardiff); minimal inflammation (malignant appearance, benign appearance); greater inflammation (malignant appearance, benign appearance); and survival (yes, no). Table 2.1 shows the numbers in each combination of the categories of these observations, set out in a typical contingency table fashion. Extra summaries can be provided by adding to the table *margins* containing totals over subsets of categories. For example, an extra column could be added at the right-hand side of the table, containing the row totals. This would show the total number of survivors and non-survivors for each category of age and centre. Similar margins can be added for other subgroups of categories, but adding these margins leads to rapid increase in the size of the table so has not been attempted in Table 2.1.

A summary table of this form will often highlight apparent trends or associations between measurements that can then be followed up using some of the techniques to be discussed in Chapters 7 and 8. For example, in the table above we may be interested in comparing survival rates for the different age groups and between the different diagnostic centres, or determining whether the survival patterns differ between age groups for the

Table 2.1.   Three-year survival of breast cancer patients

Centre	Age	Survival	Minimal inflammation		Greater inflammation	
			Malignant	Benign	Malignant	Benign
Tokyo	Under 50	No	9	7	4	3
		Yes	26	68	25	9
	50−69	No	9	9	11	2
		Yes	20	46	18	5
	70+	No	2	3	1	0
		Yes	1	6	5	1
Boston	Under 50	No	6	7	6	0
		Yes	11	24	4	0
	50−69	No	8	20	3	2
		Yes	18	58	10	3
	70+	No	9	18	3	0
		Yes	15	26	1	1
Cardiff	Under 50	No	16	7	3	0
		Yes	16	20	8	1
	50−69	No	14	12	3	0
		Yes	27	39	10	4
	70+	No	3	7	3	0
		Yes	12	11	4	1

different diagnostic centres. Also, there seems to be a different survival pattern with respect to the two categories of inflammation for the two different types of inflammation: is this a real effect, or is it just due to chance fluctuations caused by sampling? We will consider such questions again in later chapters.

Numerical measurements: summary statistics

In order to provide a concrete example on which to illustrate these quantities, consider a typical situation in industrial statistics. A car manufacturer, concerned about quality control, randomly samples sets of five camshafts produced during each of the four shifts worked each day. Various measurements are taken on each camshaft, one of them being the camshaft length. The lengths in millimetres of the 100 camshafts sampled during one week are shown in Table 2.2.

Even though this is not a large sample by present-day standards, nevertheless, there are far too many numbers here for easy assimilation by eye and some condensation is needed. Can these 100 numbers be replaced by just a few well-chosen summary measures from which the main features of the whole sample can be gleaned? If it can, what should be the nature of these summary measures?

If a single summary measure is sought in place of the whole collection of numbers then arguably the most important value is the position of the 'centre' of the data, as this value focuses attention on the approximate 'location' of the data. Unsurprisingly, statistical terminology for such a value is either a measure of centrality or a measure of location. More surprisingly (perhaps), there are several competing measures, each with its own advantages and disadvantages.

An obvious measure is the straightforward arithmetic average, given by adding up all the sample values and dividing by the number of these values.

Table 2.2. Camshaft lengths in millimetres

601.4	601.6	598.0	601.4	599.4	600.0	600.2	601.2	598.4	599.0
601.2	601.0	600.8	597.6	601.6	599.4	601.2	598.4	599.2	598.8
601.4	599.0	601.0	601.6	601.4	601.4	598.8	601.4	598.4	601.6
598.8	601.2	599.6	601.2	598.2	598.8	597.8	598.2	598.2	598.2
601.2	600.0	598.8	599.4	597.2	600.8	600.6	599.6	599.4	598.0
600.8	597.8	599.2	599.2	600.6	598.0	598.0	598.8	601.0	600.8
598.8	599.4	601.0	598.8	599.6	599.0	600.4	598.4	602.2	601.0
601.4	601.0	601.2	601.4	601.8	601.6	601.0	600.2	599.0	601.2
601.2	601.2	601.0	601.0	601.4	601.4	598.8	598.8	598.8	598.2
601.8	601.0	601.4	601.4	599.0	601.4	601.8	601.6	601.2	601.2

This is usually termed the *mean* of the sample. For the camshaft data the sum of the values is 60,014.2, so the mean is the sum divided by 100, that is, 600.142. The mean is easily calculated and does indeed point at the centre of the data. However, an unsatisfactory feature is that it is very influenced by extreme values. If, for example, one of the camshaft readings had been mis-reported—say by the omission of the decimal point (a very easy mistake to make when copying lists of figures)—then the mean would be changed dramatically. Thus if the last value 601.2 in the list above were recorded as 6012, then the mean would change from 600.142 to 654.25! Of course, data should always be checked very carefully before any analysis and such errors should be trapped, but this undue influence of extreme values can make extrapolations from a mean rather dangerous in quite ordinary situations. For example, in a test of lifetimes of batteries, out of a sample of 10 batteries it might be found that 9 of them have lifetimes of between 12 and 30 h but one exceptional battery has a lifetime of 120 h. If the first 9 have an average lifetime of 21 h, then the average of all 10 is 30.9 h—greater than the actual lifetimes of 9 out of 10 in the sample. This type of situation is not uncommon, and in such circumstances a mean may be a misleading summary.

So what alternative is there? It can be argued that a truer impression of the centre of the data is obtained by arranging the values in order (it does not matter whether it is increasing order or decreasing order), and then finding a value x such that half of the values are less than x and half of the values are greater than x. The value x found in this way is called the *median* of the sample. There are various technical details associated with the calculation, such as how to cope with repeated values or large gaps between values, but they need not concern us here; it is just the concept that is important. Effecting this process on the camshaft data, we find the median to be 600.6. Note that, as a by-product of the method of obtaining it, the median does not suffer from sensitivity to a few extreme values. Thus if the value 601.2 is indeed misreported as 6012, the median will remain at 600.6. This is because 601.2 belongs to the half of the sample greater than 600.6, and changing it to 6012 does not affect this positioning so will not affect the median. Likewise, the median of the battery lifetimes will not change whether the longest recorded lifetime is 30 h or 120 h. A more extreme displacement occurs when an error is made in one or two observations in the 'wrong half' of the sample—for example, if the value 598.8 below the median is misreported as 5988 (above the median)—so that the value of the median shifts upwards, but this change will only be substantial if there are big differences between values in the sample. Of course, there is a price to pay for this stability, and this price is that the actual values of the individuals in the sample have not been used—only their relative positioning plus the values of the 'central' individuals.

But the mean and the median are not the only plausible measures of centrality, and cases can be made for various other measures. Probably the only other likely measure that readers may encounter is the *mode* of the sample, which is the most frequently occurring value. For the camshaft data, 601.4 occurs 13 times, 601.2 12 times, 598.8 11 times, 601.0 10 times and all other values have single-figure frequencies of occurrence. So the mode of the sample is 601.4, but 598.8, 601.0 and 601.2 are all near-modal values also.

All three of these measures are valid and useful measures of centrality or location, and choice of one or other of them may depend on the context of the problem. Indeed, in any single case, each measure may throw different light on matters. Earlier we commented on the problems in using a simple random sample to estimate the average monthly wage bill of a large company containing 50 top executives, 150 middle managers and 800 administrative and clerical workers. Consider the problem of quoting a 'centre' of the wage bill. The owners of the company might wish to quote the mean of all 1000 employees, as this will include the top wage earners and hence show the company in best light with regard to rewards offered to staff. The unions, on the other hand will prefer to quote the mode in order to demonstrate that the majority of staff are poorly paid. In order to present a more balanced picture, avoiding the extreme influence of the highly paid top executives but, nevertheless, acknowledging the large group of administrative and clerical workers, the median would probably be the best measure.

Whichever measure of centrality is chosen in a particular case, it is evident that reducing a set of sample values to a single summary value leads to oversimplification. Many samples that differ in their general features can end up with the same mean or median, and will then be mistakenly treated as similar. So we need at least one other summary measure to help distinguish them. Of the various possible features of a sample, the next most important one after its centre is the *spread* of its values. Clearly, if a set of sample values is tightly bunched about its centre then we will have more confidence in making statements about the measurements, or predicting future values, than if the sample is widely dispersed about its centre. So a summary measure of spread (with synonyms 'variability' and 'dispersion') is our next objective.

The simplest measure of spread is the *range* of the sample, namely the difference between the largest and smallest observations. However, this may sometimes be an unnecessarily conservative measure. Accepting that the largest and smallest observations can often be 'unusual', 'extreme' or even 'unrepresentative', we may want to exclude some of them from the calculation. If we are using the median as the measure of centrality, then we can extend the argument that led to the value x as the value dividing the range of the measurements into two equal (50%) portions and find the two values y and z which divide the range into 25%/75% and 75%/25% portions. Thus 25% of the sample values are less than y, and 75% of the sample values are less than z. The two values y and z are known as the *quartiles* of the sample, and the difference $z - y$ is the *interquartile range*. This provides a good measure of spread to go alongside the median, as it gives the range of the 'middle' 50% of the data and puts the value of the median into its relevant context. For the camshaft data, the two quartiles are 599.63 and 601.14, respectively, so the interquartile range is 1.51.

If we recognise that the interquartile range is a measure of spread based on the same building blocks that constructed the median, then when we turn to the mean we need a measure of spread alongside it that uses the concept of averaging values in some way. The common measures here are a little more complicated, but can be intuitively justified as measuring the average distance away from the mean of the sample observations. For technical reasons it is preferable to consider the squared differences between each observation and the sample mean, and the (slightly adjusted) average of these squared differences is the *variance* of the sample. Clearly, the more dispersed the sample the bigger will be some of the squared differences, and hence the bigger will be the variance. The benefit of squaring differences is that as discrepancies between an observation and the mean become larger, so the squared discrepancies become progressively more heavily emphasised. Thus the variance is sensitive to the presence of extreme observations. However, since the variance is a measure of *squared* differences it cannot be directly compared with the mean, the two quantities having different scales. For direct comparison to be made possible we must take the square root of the variance, thereby equalising the scales. This square root is the *standard deviation* of the sample. Readers who have not come across this term before, but who are familiar with other scientific terms will perhaps recognise it as the root mean square error used in physical sciences. For the camshaft data, the variance is 1.796 and the standard deviation is the square root of this, or 1.34.

So we now have measures of spread of the sample to go alongside the median and mean. Are two measures sufficient to describe a sample? In many cases it turns out that they are, but a few common higher-order measures have been defined in order to cope with cases when they are not. One is a measure of *skewness*, which is designed to indicate to what extent the sample departs from symmetry (i.e. to what extent sample values are more bunched on one side of the mean and more spread out on the other), and another is a measure of *kurtosis*, which is designed to indicate to what extent the sample values are heavily concentrated round the mean as opposed to being more evenly spread across the whole range. The coefficient of skewness of a set of data is the (slightly adjusted) average of the third powers of the differences between the observations and the sample mean, divided by the third power of the standard deviation. A negative value of the coefficient indicates that the observations are more spread out below the mean than above it, a positive value indicates the opposite, and a value close to zero suggests a symmetric sample. The coefficient of kurtosis is the (slightly) adjusted average of the fourth powers of the differences between the observations and the sample mean, divided by the fourth power of the standard deviation. The larger the value of this coefficient, the more evenly spread are the sample observations. The 'slight adjustment' in the calculations of variance, skewness and kurtosis arises from division by one less than the sample size during the averaging process, while the division by a power of the standard deviation in the calculation of skewness and kurtosis is simply a normalisation mechanism.

As a final point, it is worth giving a few indications of the interrelationships between these measures. If the sample is reasonably symmetric then the mean and median will be similar in value (as with the camshaft data, for example), but if there is a big difference between these two measures then that is an indication of skewness of the sample values. If the sample is reasonably symmetric and does not have unexpectedly extreme observations, then an interval formed by taking a width of two standard deviations on either side of the mean will usually contain about 95% of the sample values, while an interval of three standard deviations on either side of the mean will usually contain all except one or two observations. The standard deviation, in particular, will play a central role in later discussions.

Summary distributions

If reduction of the sample values to just a few summary measures is considered to be too drastic, then how can we simplify the set of values in such a way as to convey their general features while at the same time reducing the amount of detail as well as the space required to present the data? In the case of the camshaft data, we notice that some reduction is immediately possible, because the 100 numbers consist of fewer distinct numbers along with repetition of some of the values. We have already noted above that the values 601.4, 601.2, 598.8 and 601.0 occur 13, 12, 11 and 10 times each respectively. If we go through the list completely, we find that there are just 23 distinct values and the number of repetitions ranges from 13 for 601.4 to 1 for each of 600.2, 600.4 and 600.6. So we could reduce the number of values written down from 100 to 46, namely the 23 distinct values and the *frequency* (i.e. number of repetitions) of each, without losing any of the detail.

However, such a reduction is arguably not very good, because a list of 23 separate values with frequencies alongside is still rather lengthy. Moreover, the camshaft lengths are slightly unusual, in that they have evidently been recorded only to the nearest 0.2 of a millimetre. Most measurements in practice would be recorded to rather more detail, for example, lengths such as 602.386, 605.429, 597.584 and so on. With data of such accuracy, the occurrence of repetitions of values is highly unlikely. However, irrespective of the accuracy or type of measurement, genuine simplification is achieved by *grouping* the measurements and writing down the frequency of occurrence of each group. For example, the camshaft lengths can be grouped into intervals 1 mm wide and the frequencies of each group determined. This will give a compact summarisation of the data, as shown in Table 2.3.

Note here that the symbol '596.5 $\leq x$' means '596.5 is less than or equal to x', while '$x < 597.5$' means 'x is less than 597.5'. So '596.5 $\leq x < 597.5$' indicates camshafts whose lengths (x) are greater than or equal to 596.5 but less than 597.5, and the entry in the column headed 'Frequency' shows there is just one camshaft satisfying this condition. Similar interpretation holds for the other lines in the table. The extra column headed 'Relative frequency' expresses the frequencies as proportions of the total, and the values in this column are simply obtained by dividing the corresponding values in the 'Frequency' column by the total of that column (100 in this example). Relative frequencies are useful if data sets containing different numbers of values are to be compared, as direct comparison of frequencies is difficult in those cases. Also, a relative frequency indicates the probability of randomly drawing an individual within the required interval from the sample, and such probabilities will be taken up again in the next chapter.

Thus the table below, known as a *frequency distribution*, gives a good summary in compact form of the main features of the whole data set. There are two main considerations when summarising data in this way. The first is to ensure that the intervals are unambiguous (i.e. do not overlap) and cover the whole range of values. This is why we have $\leq$ at the start of each interval but $<$ at the end—so that values such as 597.5 would unambiguously belong to the second interval, and not be a candidate for both first and second intervals. (Of course, since the camshaft lengths are measured to the nearest 0.2 of a millimetre, values such as 597.5 do not occur; but tables should always be presented in a fashion that is clearly unambiguous to the reader, even if the underlying data values have not been seen). The second consideration is to choose an appropriate number of intervals: few enough

Table 2.3. Frequency distribution of camshaft lengths

Camshaft length (x)	Frequency	Relative frequency
596.5 $\leq x < 597.5$	1	0.01
597.5 $\leq x < 598.5$	16	0.16
598.5 $\leq x < 599.5$	24	0.24
599.5 $\leq x < 600.5$	8	0.08
600.5 $\leq x < 601.5$	41	0.41
601.5 $\leq x < 602.5$	10	0.10
Total	100	1.00

to make for simplification of the data, but sufficiently many to ensure that the detail in the data is not lost. Usually somewhere between about 6 and 12 intervals should be appropriate. Here we have gone for the lower value, partly because intervals of width 1 mm are natural ones to take, and partly because these intervals give a good impression of the shape of the data (rising to an early peak, then dropping but rising again to a larger peak before tailing off).

Unfortunately, the *quid pro quo* of summarisation is that some detail is always lost. We no longer have the individual values, but can only say how many camshafts have lengths within an interval of 1 mm. This is not a problem if the summary is for presentation purposes only. But if we then want to conduct any calculations from the summary table, we have to make some assumptions and our calculations will only yield approximate values. The standard assumption for frequency tables is that the individuals in a particular interval have values evenly spread throughout that interval. For example, we assume that the 41 camshafts in the fifth interval have lengths evenly spread between 600.5 and 601.5. For purposes of calculating quantities like the mean or variance, this is tantamount to assuming that all individuals in an interval have value in the middle of that interval, for example all 41 in the fifth interval have value 601.0. So the mean is calculated by multiplying the midpoint of each interval by the frequency in that interval, adding up the resultant products and dividing by the total number of individuals. Notice that this is exactly equivalent to simply multiplying each midpoint by the corresponding *relative frequency* and adding up the resultant products. With this assumption, we would estimate the mean length of the camshafts as 600.02—not far off the 'true' value of 600.142. The variance is likewise calculated by multiplying the squared difference between each interval midpoint and overall mean by the relative

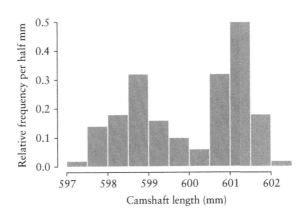

Figure 2.1.   Histogram of the camshaft relative frequencies.

frequency in that interval, and adding the resultant products. For the camshaft data this gives an estimated variance of 1.798 as opposed to the 'true' value of 1.796, again a very close approximation. Calculation of the coefficients of skewness and kurtosis proceeds exactly as for the variance, but using the third and fourth powers respectively of the differences between the interval midpoint and the overall mean in place of the squared differences. This grouped frequency distribution approach has important implications for population models, as considered in the next chapter.

A final point to make is that pictorial representations of data are extremely useful for showing general features, but also suffer from inability to show finer detail. The most common representation of grouped frequency data is the *histogram*. On such a diagram, the measurement of interest (x) is marked on the horizontal axis, and the frequencies or relative frequencies in each interval are represented by rectangles with bases on the appropriate x intervals and *areas* equal to the corresponding frequencies or relative frequencies. Thus if the latter are being represented, then the total area is equal to 1 and the vertical scale is the relative frequency *per unit class interval*. Figure 2.1 shows a histogram of the relative frequencies of the camshaft data, grouped now into 11 half-millimetre intervals in order to highlight greater detail.

3 Population Models

Introduction

We have seen that the population is generally the focus of any statistical investigation. It is the collection of all possible individuals that we think are relevant to our study, and the purpose of most statistical investigations is to obtain information about some specific measurement of interest in this population. Usually, the population is too large for us to be able to observe every individual in it, so we will hardly ever be in a position to know all we can about this measurement of interest. Suppose for the moment, however, that all members of the population *could* be observed—so that we *could* obtain information about the distribution of all values of our measurement of interest in the population. Such information would tell us about the relative frequencies of values in different intervals, which would in turn tell us about the chances of getting particular values of the measurement when drawing individuals at random from the population. This would in turn be very valuable in various ways: in predicting future values, in guiding future actions, or in making decisions on the basis of these chances. For example, if we were a clothing manufacturer and we knew exactly the frequencies of individuals in the population taking each size of suit, then we could manufacture the right proportion of suits of each size in order to match supply to demand exactly.

Unfortunately, most populations of interest are either many times too large to permit such exhaustive measurement, or indeed they are infinite so we will never be able to measure all individuals in them. The camshaft data constitute a good example—the population consists of all camshafts manufactured by the company, but this will never be complete during the lifetime of the company: every new camshaft manufactured has the potential to be different from all previous ones, so the end of the population can never be reached. What can we do in such circumstances? The course of action usually taken is to *model* the behaviour of the measurement in the population, either from knowledge of the nature of the measurement or from knowledge of the mechanism that generates it. Most commonly, the model will be a mathematical expression involving one or two unknown quantities, known as *parameters*, such that if numerical values are substituted for these parameters then complete information about population probabilities is provided. Thus any probabilities that we might be interested in, such as the ones mentioned above, can be calculated. The accuracy with which we calculate these probabilities is of course affected by the values that we use in place of the model parameters, and the choice of these values is aided by sampling from the population. This takes us into the province of *statistical inference*, namely the process of extrapolating from the sample to the population, which is broadly the subject matter of the first half of this book. In this chapter we lay the groundwork, by considering the modelling of populations in a little more detail.

Probability distributions

In Chapter 1, when discussing the probability of winning in the game of craps we denoted the sum of the scores on the first roll of the two dice by B, and then by using the 'equally likely' approach to probability we produced Table 3.1 of possible values of B and their associated probabilities.

This is an example of a *probability distribution*, so named because it shows how the probabilities of all possible values of some quantity (here B) are distributed. Although this probability distribution was derived by mathematical argument based on the equally likely approach to probability, something very close to it could also be obtained empirically by throwing a pair of dice a large number of times and listing the relative frequency of occurrence of each value of B. The more times we threw the dice, the closer in general would be the relative frequencies to the given probability values. With only 50 or 100 throws the two sets might differ considerably, because of random chances and small numbers of trials. However, once we had thrown the dice 1000, 5000 or 10,000 times, the differences between the two sets would be very small. The probabilities below are essentially just the relative frequencies of the whole population of potential throws of a pair of dice, and as the sample gets closer and closer to the population so the two sets of relative frequencies match more and more closely. Here the problem is that the 'population' is infinite, as we can simply carry on throwing the dice forever; but once the number of throws becomes 'large enough', there should be no practical differences between the sets. Mathematicians say that

Table 3.1. Probability distribution of score on throwing two dice

B	2	3	4	5	6	7	8	9	10	11	12
$P(B)$	$\frac{1}{36}$	$\frac{2}{36}$	$\frac{3}{36}$	$\frac{4}{36}$	$\frac{5}{36}$	$\frac{6}{36}$	$\frac{5}{36}$	$\frac{4}{36}$	$\frac{3}{36}$	$\frac{2}{36}$	$\frac{1}{36}$

'in the limit' (i.e. as the number of throws becomes impossibly large!) the distribution of sample relative frequencies becomes equal to the probability distribution.

Using this limiting relative frequency argument, we see that we can calculate summary values such as mean, variance and standard deviation, coefficient of skewness, and coefficient of kurtosis for probability distributions of quantities in the same way as we did for relative frequency distributions at the end of the previous chapter. The mean is obtained by multiplying each possible value of the quantity (*B* above) by its corresponding probability (i.e. population relative frequency) and adding the resulting products. The variance is calculated by multiplying the squared difference between each value and mean by the corresponding probability and adding the resulting products, and the standard deviation is, as usual, the square root of the variance. For the probability distribution in Table 3.1 we thus find that the mean is 7.0, the variance is 5.833, and hence the standard deviation is 2.415. The mean of the probability distribution of a quantity is often termed the *expected value* of that quantity, it being the value we would expect as the average over a large number of realisations of the process. So the expected value of the score on throwing two dice is 7.0. If needed, higher order moments such as skewness or kurtosis can be calculated in the same way as the variance, but replacing the squares of the differences between each value and the mean by their third or fourth powers respectively.

The above example is one in which the probability distribution can be exactly determined by mathematical calculation, but this is not usually possible in most practical studies. In these cases we have to resort to making some *assumptions* about the nature of the probability distribution, which means that we formulate a *model* of the population. This model is generally a mathematical expression, or *function*, from which any necessary probabilities can be calculated. A mathematical function is simply an equation enabling the value of one quantity to be found if the values of other quantities are supplied. For example, if the function is $y = x^2$ then supplying values 1, 2, 3 in turn for x provides values 1, 4 and 9 in turn for y. For modelling a probability distribution, the value of x would be the value of the quantity of interest (e.g. the value of B in the example above) while the resulting value of y would be the corresponding probability of observing this value of x. We would usually look for a function that produces probabilities which are in reasonable accord with any data collected from the population in question.

As might be expected, there are many potential mathematical functions that could be used to model the population, and choosing a good one can be something of an art as well as a science. Note that we say a 'good' one rather than the 'right' one, because in any given situation there might be several models that are equally good, in the sense that they all satisfy the necessary assumptions and constraints, and all accord reasonably well with the observed data. Indeed, it can never be established unequivocally whether a particular model is the 'right' one, because logically to do so we would need to have measured the whole population! Of course, even though we may not know what the right model is, we want to select one that is appropriate and avoid any that are inappropriate. Different models have different characteristics, so in order to avoid making a bad choice we need to match at least some of the primary characteristics of model and data. The main such characteristics are *type of measurement* and *shape of distribution*, but even when we have matched on these characteristics we have considerable choice of available models. At that stage, we perhaps need to introduce outside knowledge of the substantive area of the data in order to make our choice, or alternatively we might appeal to the *data generation mechanism* in order to provide some finer detail. We therefore consider each of these aspects in turn, and survey the most common choices of probability models in practical use.

Type of measurement

This is probably the most important characteristic of the population to establish at the outset, and if data and model do not match on this characteristic then any resulting statistical analysis can go badly wrong. There is a clear distinction between models for *discrete* measurements, namely ones that take only distinct possible values such as integers 1, 2, 3, and so on, and *continuous* measurements, namely ones that can take any value in a given range as, for example, 34.672 or -25.44. The reason why this distinction exists is because we can always attach probabilities directly to the individual values of a discrete measurement, as with the probabilities above for the scores obtained on throwing two dice, but for a continuous measurement the probabilities have to be associated with intervals of values, as with the camshaft data in the previous chapter.

This may seem to be a trivial distinction, but mathematical complications arise since a model is a mathematical expression from which these probabilities are computed. Thus for a discrete measurement, the model is fully specified by a mathematical expression which yields a probability value whenever we provide it with the value of the discrete measurement. So the probability is obtained from the model simply by direct substitution of discrete value into the formula. For example, a simple mathematical expression for the probabilities of obtaining a score B with a throw of two dice is:

$$P(B) = \frac{B-1}{36} \text{ whenever } B \text{ is less than or equal to 7, and}$$
$$P(B) = \frac{13-B}{36} \text{ otherwise.}$$

It is easy to check that all the probabilities in Table 3.1 are obtained on substitution of the appropriate value of B into this expression.

Summaries such as mean (i.e. expected value) and variance are then calculated from these mathematical expressions in the way described earlier. Such summary values for populations have traditionally been denoted by Greek letters: μ (pronounced 'mu') is generally used to denote the population mean, σ ('sigma') the population standard deviation, and hence σ^2 ('sigma squared') the population variance. Thus the expected value μ of B is given by adding the values of $B \times P(B)$ over all the possible values of B, and likewise the variance σ^2 of B is given by adding the values of $(B-\mu)^2 \times P(B)$ over all the possible values of B. As regards higher order moments, β ('beta') generally denotes the skewness and κ ('kappa') the kurtosis.

However, for a continuous measurement we could divide the range up into all sorts of intervals; for example, the camshaft data were grouped into intervals of width 1 mm, but we could just as well have grouped them into intervals of 1.2 or 1.5 or 2 mm. If we want to use a population model for any type of analysis of the data, we need to be able to associate probability values with intervals *whatever the width of these intervals*. Consequently, it is not possible to provide a single mathematical expression that will directly generate by means of a simple substitution the probabilities of individuals having values in given intervals. Instead, the solution is to provide a single mathematical expression, called a *probability density function* and denoted by $f(x)$, which when evaluated at different values of x over the whole range of the continuous measurement traces out a curve of some form on graph paper. This curve is in effect the outline of the histogram of relative frequencies in the whole population, and has the property that the probability of obtaining a value of x in any interval is the area of that interval under the curve. We can therefore calculate the probability associated with any interval by finding this area, so at this point a fundamental difference becomes evident between the treatment of discrete and continuous measurements. Probability calculations relating to discrete variables can usually be carried out with the aid of simple algebra and arithmetic, but to find areas under curves we need to employ more complicated methods, generally involving integral calculus. Readers unfamiliar with these more advanced mathematical techniques need not fear, as we will not be going into any details, but it is worth bearing in mind that they invariably underlie the calculations whenever we deal with continuous variables. Of course, nowadays there are many sophisticated computer programs or even functions on pocket calculators that we can call upon to carry out the necessary calculations, whereas in the past the only assistance offered to the user were rather cumbersome tables of areas for different mathematical functions.

Extending the above connection between probabilities for discrete measurements and probability density functions for continuous measurements, and continuing with the concept of areas under functions in the latter case, the expected value μ of a continuous measurement is obtained by finding the total area under the new function $xf(x)$, and the variance of a continuous

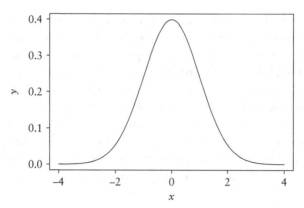

Figure 3.1. Probability density function of the standard normal distribution.

function by the total area under the second new function $(x - \mu)^2 f(x)$. These calculations of course again involve the use of integral calculus.

There are many different mathematical expressions that can be used as probability density functions. The only requirements for an expression to qualify are that its value must nowhere be negative (as negative probabilities are impossible), and that the total area under its curve must equal 1 (as the probability of obtaining an outcome *somewhere* in its range is 1). The most common such function, which we will meet frequently throughout the book, is the (standard) normal density function—so we show a plot of this density function in Figure 3.1. Values of x, the measurement of interest, are shown along the horizontal axis, and they trace out the bell-shaped density function curve y. Since the probability of any interval of values of x is given by the area under this curve for that interval, it is evident that probabilities of intervals will be highest in the middle of the range shown, and will steadily decrease for intervals moving away from the middle in either direction until they drop to virtually zero at the ends of the range shown.

Shape of distribution

Having decided on what type of measurement we are interested in modelling, the next consideration when narrowing the set of possible models is to decide (roughly) what the form of the population distribution is likely to be. In other words, are the maximum probabilities likely to occur somewhere in the middle of the range of values of the measurement (like in Table 3.1 or Figure 3.1), or at one or other end of the range? Distributions of quantities such as clothing sizes are likely to be as in Figure 3.1: there are very many people of 'average' height and weight, progressively fewer people in more extreme sizes (either smaller or larger than average), and very few people at the ends of a reasonable scale. On the other hand, a manufacturer of electrical light bulbs might be interested in the distribution of their lifetimes, and it is a well-known phenomenon that the risk of a bulb 'blowing' is highest with the power surge when it is switched on. So a more reasonable model might be one in which the probability density function starts with quite a high value for low usage times, drops to a fairly low value after a reasonable 'start-up' time, but then rises again when the 'age-effect' kicks in and the old bulb is likely to wear out. Such a density function is said to be 'U-shaped', in contrast to the bell-shaped curve of Figure 3.1. Different mathematical functions are available as models for these various eventualities; so choosing an appropriate function from this set requires some appreciation of the type of shape represented by each one.

Some common probability models

We list here the half-dozen most common probability models, three for discrete measurements and three for continuous measurements, as between them these half-dozen models probably account for most of the modelling situations the reader is likely to encounter in practice. There are of course many other possibilities, but most are more complicated than the following and hence are usually employed in more specialised circumstances. Each of the models involves one or more parameters that have to be substituted by a specific value for probabilities to be calculated, and most of the models can generate differing shapes of distribution depending on the parameter values chosen. As with the earlier population summary statistics, such parameters are generally denoted by Greek letters so we will indicate the pronunciation of each new letter as it occurs. We will discuss possible motivations for the models, situations in which they might be applicable, and considerations in the choice of parameter values subsequently. First, however, we simply give the name of each model, specify the range of values of the measurement that are supported by it, the permitted range of values of the parameter or parameters, and any available indication about the shape of the distribution that the model generates. We also give the mathematical formula for either the discrete probabilities or the continuous probability density function in each case, but stress that these formulae are given for completeness only and can be ignored by less mathematically inclined readers. The three models for discrete measurements come first, followed by the three models for continuous measurements.

The binomial distribution

This model is appropriate when the discrete measurement can take integer values between 0 and some maximum value n, so is ideally suited when counting, say, the number of individuals out of a sample of size n that possess some attribute of interest (e.g. the number of people in a sample voting for a particular political party). Thus n is a parameter that has to be specified, and there is a second parameter θ ('theta') that is permitted to have any (continuous) value between 0 and 1. Depending on the value chosen for θ any shape of distribution can be generated, from one having maximum probability near zero and then gradually tailing off, through one having its maximum somewhere in the middle of the range and probabilities declining fairly symmetrically on either side of the maximum, to one where the probabilities are banked up with maximum somewhere near n. The probability of the measurement taking value i with this model is $(n!/i!(n-i)!)\theta^i(1-\theta)^{n-i}$, where $n!$ is shorthand for the product of all integers between 1 and n, that is, $1 \times 2 \times 3 \times \cdots \times n$. The expected value and variance for this distribution are $n\theta$ and $n\theta(1-\theta)$ respectively.

The Poisson distribution

This model is appropriate when the discrete measurement can take any positive integer value or zero, so is ideally suited to situations in which some unrestricted counting process is being conducted (e.g. numbers of bacteria on different slides in a microbiological laboratory). It has one parameter, λ ('lambda'), which can take any (continuous) positive value. If this value is less than 1 then the maximum probability occurs when the measurement is 0 and the probabilities gradually decline with increasing measurement value, but if the parameter value is greater than 1 then the probability distribution first rises to a maximum and then falls away to zero. The speed of rise and decline is governed by the value of λ. The probability of the measurement taking value i with this model is $(\lambda^i/i!)e^{-\lambda}$, where e is the transcendental number equal to 2.7183 (approximately) and e^x is termed the exponential function of x. The expected value and variance of this distribution are both equal to λ. Note that $e^{-\lambda}$ can also be written in the form $\exp(-\lambda)$.

The geometric distribution

This model is essentially a mixture of the two preceding ones: it is appropriate when the discrete measurement can take any positive integer value or zero, but its one parameter θ is restricted to have a (continuous) value lying between 0 and 1. Also, the shape of the distribution is always one of gradual decline from a maximum at measurement value 0. So this model is considerably less flexible than the two previous ones, but is nevertheless very useful in certain circumstances. The probability of the measurement taking value i with this model is $(1-\theta)^i\theta$; the expected value and variance are equal to $(1-\theta)/\theta$ and $(1-\theta)/\theta^2$, respectively.

The normal distribution

The first model for continuous measurements that we consider is probably the most used and familiar distribution in all statistics. It is associated with Carl Friedrich Gauss, a prolific German mathematician who lived in the last quarter of the eighteenth and the first half of the nineteenth centuries, and who established the beginnings of the statistical method. The distribution is consequently still frequently referred to as the Gaussian distribution. It can be used to model a continuous measurement that can take any value, whether positive or negative. The probability density function has two parameters: μ, which can take any continuous value whether positive or negative, and σ, which is restricted to positive (continuous) values. Whatever the values of the parameters, the probability density function is always symmetric and bell-shaped with its maximum in the centre. The value of μ determines the position of this centre, while the value of σ determines the width of the central portion of the bell. In fact, in a population of individuals following a normal distribution, μ is the mean of the measurements while σ is their standard deviation. The measurements can be *standardised* by subtracting μ from each one and dividing the result by σ. This produces a set of values that have a *standard normal distribution*, namely one whose mean is 0 and whose standard deviation is 1. This is a very useful procedure, because in this way we can reduce every normal distribution to a single case if we know the parameter values, so probabilities for any normal distribution can be found from a single table of probabilities. A graph of the probability density function for the standard normal distribution has already been encountered, in Figure 3.1. More generally, the mathematical formula for the probability density function of a normal distribution having mean μ and standard deviation σ is

$$f(x) = \frac{1}{\sigma\sqrt{2\pi}} \exp\left\{\frac{-1}{2\sigma^2}(x - \mu)^2\right\},$$

where π is the familiar transcendental number 'pi' with value 3.1416 (to four decimal places) and x is the value of the measurement at which the density function is calculated. Thus Figure 3.1 shows a plot of $f(x)$ when $\mu = 0$ and $\sigma = 1$. Varying the value of μ changes the position of the centre of the curve, while varying the value of σ varies its width. Figure 3.2 shows two more normal distributions, both having $\mu = 0$ but one with $\sigma = 2$ (solid line) and the other with $\sigma = 4$ (dotted line). As σ increases, the centre of the curve becomes progressively lower, while the curve becomes progressively wider and the decrease in height from the centre outwards becomes progressively shallower. Consequently, the probability of obtaining a value in any fixed interval on one side of the mean, that is, the area under the curve of any fixed interval on one side of the centre, progressively increases. Empirical observation, underpinned by mathematical

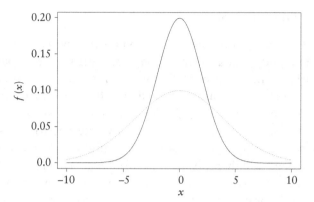

Figure 3.2. Density functions of two normal distributions.

results known as central limit theorems, has established that the normal distribution is a very suitable model for many measurements made in practical studies. In particular, measurements such as lengths, weights, areas, volumes and suchlike can usually be modelled by normal distributions.

The Weibull and related distributions

Although the normal distribution provides a good model in many situations, there are nevertheless many other situations in which it is patently inadequate. In particular, it will not be suitable for any data sets that show marked asymmetry, and a common measurement that leads to data of this type is time. Examples occur in survival analysis in medical research, where interest focuses on survival times of patients under treatment for serious diseases, or in reliability analysis in engineering, where interest focuses on times to breakdown of machine components, or in environmental studies where interest focuses on time to occurrence of natural events such as floods or earthquakes. Data from such studies have several very non-normal features, for example, the possibility of very asymmetric distributions that cater for high probabilities of short times, and non-negligible chances of extremely long times. The Weibull distribution provides a very flexible model for such situations. It has two parameters, a and b, both of which can take any continuous positive value, and a variety of shapes of distribution can be generated with judicious choice of values for these parameters. The formula for the probability density function is

$$f(x) = abx^{b-1}\exp\{-ax^b\}.$$

If the value 1 is chosen for b, then the density function reduces to $f(x) = ae^{-ax}$, a very simple form. The resulting distribution is known as the *exponential distribution*, and this is also a popular model in its own right. A related but slightly different form, in which the x^b in the exponential term is replaced

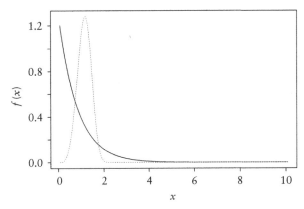

Figure 3.3. Density functions of two Weibull distributions.

by just x and the leading constants a and b are amended appropriately, is known as the *gamma* distribution and is also a popular model for lifetime data. To illustrate some possible shapes the Weibull distribution can take, Figure 3.3 shows the density functions when $a = 1.2$, $b = 1$ (solid line) and when $a = 0.5$, $b = 4$ (dotted line).

The beta distribution

A final situation not covered by the above two models for continuous measurements is when the measurement value is restricted to the range $0-1$. Such a situation is frequently encountered, as *proportions* fall in this range and interest is often centred on proportions. Examples include quality assessment, where attention may focus on proportions of certain chemicals in different food products, or geology, where proportions of different minerals in rock samples might be of interest, or social studies, where proportions of individuals in particular social categories might be compared across different towns or countries. A useful model for such situations is provided by the beta distribution, which has two parameters a and b each of which can take any positive continuous value. By judicious choice of these parameter values any sort of shape can be generated for the probability density function, including a shape having a peak at one end of the range ('J-shaped') or peaks at both ends of the range ('U-shaped') as well as the more traditional peak in the middle of the range. This model is thus a very flexible one, but its probability density function has a slightly more awkward formula than the other ones considered so far. It is

$$f(x) = \frac{1}{B(a,\, b)} x^{a-1} (1 - x)^{b-1},$$

where the term $B(a,\, b)$ is known as the *beta function* whose value for particular a and b has to be found either from a table or by computer calculation. This feature makes the beta distribution slightly more troublesome

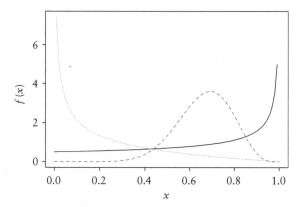

Figure 3.4. Density functions of three Beta distributions.

to handle, but it is nevertheless used quite often in practice. Again, to illustrate the variety of shapes that can be generated with the beta distribution, Figure 3.4 has three examples: when $a = 1$, $b = 0.5$ (solid line), when $a = 0.5$, $b = 2$ (dotted line), and when $a = 10$, $b = 5$ (broken line).

Choosing a model

We have mentioned in the course of the above descriptions some of the main characteristics of each distribution, along with some of the typical situations for which they may be useful. This may be enough for us to pin down an appropriate distribution to use as a model in a given situation. For example, if we are studying some common continuous measurement such as height, weight, length, blood pressure or suchlike, then the normal distribution would be a good starting point for a model of the population. On the other hand, a proportion would clearly lend itself to modelling by a beta distribution, while for a survival time we could consider a Weibull or a gamma distribution. In some cases the issue may not be clear-cut from these considerations, and we might look first for a little more guidance. If we have some knowledge of the data generation mechanism then this might offer us some clues, and there are two such mechanisms that are particularly useful in this respect.

The first one is when the data could be assumed to arise from a series of independent 'trials' in each of which the outcome has to be one of two possibilities, conventionally denoted 'success' and 'failure', and the probability θ of obtaining a success does not change from trial to trial. Such trials are known as *Bernoulli trials* after Jacques Bernoulli, the seventeenth-century originator of many results in probability theory. It can easily be shown that the total number of successes in n Bernoulli trials follows a binomial distribution with parameters n and θ, so the binomial distribution would be the right model for such data. Practical examples abound. For instance, an opinion poll might be conducted to determine which of two political parties A and B has majority backing in a particular city. If a vote for party A is deemed to be a 'success', if it is assumed that any individual chosen at random from the city has probability θ of declaring for party A, and if people in the survey respond independently of each other, then a single person's opinion is the outcome of a Bernoulli trial and the total number of individuals declaring for party A out of a random sample of n people can be confidently modelled by a binomial distribution.

A second data generation mechanism that can provide useful guidance concerns events occurring randomly in time or space. An example here might be a study of traffic accidents in a particular city, as the starting point of a project that aims to improve road safety. One part might focus on a particular stretch of road, and record the times at which accidents occur—this is a study of events in time. Another part might record the locations of accidents across the whole city, regardless of the time at which they occur—this is a study of events in space. Of course, if we also record times of accidents in the latter case then we have a study of events in time *and* space, but this introduces much greater complexity to the modelling so let us just focus on the simpler cases. It is often reasonable in these cases to make the following assumptions: (i) the events happen at a constant rate, say λ per unit of time or space; (ii) the events occur independently of each other and (iii) not more than one event can occur at exactly the same instant (time) or position (space). These conditions define what is known as a *Poisson process*, and mathematical theory then tells us that under these conditions the number of events to occur in a given time (or given area) t has a Poisson distribution with parameter λt, and the time (or distance) between successive events has an exponential distribution with parameter λ. So these distributions would provide good models for data of these types; indeed the Poisson distribution is generally the first port of call when modelling occurrence of events in time or space, while the exponential distribution is likewise the first one for modelling inter-event times.

Estimating parameter values

The choice of probability distribution to model a population is fairly restricted to start with, and considerations such as those above help to narrow the choice down to one or two possibilities in most cases. So the choice of distribution is generally not a great problem. However, choosing the distribution is only the first step, as we then have to choose appropriate values of any parameters in this distribution, and this choice can hardly ever be made without some information from the population itself. In the absence of any firm substantive knowledge, this information usually has to come from a sample collected from the population in question.

Table 3.2. Frequency distribution of numbers of defects on 400 pieces of cloth

Number of defects	0	1	2	3	4	5	Total
Number of pieces of cloth	92	142	96	46	18	6	400

So, assuming that we have fixed on a probability model for the population and have collected a random sample of values from this population, we in effect need to *estimate* the values of the population parameters from the data in the sample. How do we go about this task?

Let us examine the sort of reasoning that we might follow in a simple example. Suppose a manufacturer is concerned about the quality of a particular type of cloth in terms of the number of defects present in a standard-sized piece of the cloth, and wants to model the distribution of numbers of defects on all standard-sized pieces of cloth in the population. This situation concerns events (i.e. defects) in space (i.e. pieces of cloth). So from the discussion above, a Poisson model is suggested for the population. Under this model, the probability of finding r defects on a randomly chosen piece of cloth is $(\lambda^r/r!)e^{-\lambda}$, for some suitable value of the parameter λ. But what should this value be? To obtain some guidance, the manufacturer collects 400 standard-sized pieces of cloth at random from the manufacturing process, and observes the results shown in Table 3.2.

The first point to note is that if the probability of finding r defects on a randomly chosen piece of cloth is as given above, then out of 400 pieces of cloth we would expect to find about $400 \times (\lambda^r/r!)e^{-\lambda}$ pieces that have r defects. So for a particular value of λ we would have a particular distribution of these expected numbers. Clearly, a 'good' choice of λ would be one whose distribution resembled the actual frequency distribution above, whereas if the two distributions were very different then that would be a poor choice of λ. So the choice of λ reduces to finding ways of aligning the observed distribution with the theoretical one obtained from the population model.

Perhaps, the simplest way of aligning the observed and theoretical distributions is to make sure that their 'main features' match exactly. We have seen earlier that the main features of a sample are represented by their mean and variance, and possibly also their skewness and kurtosis. We have also seen that it is possible to calculate mathematically the population values of these quantities for any theoretical model; some of these values have been given above for common models, and most standard textbooks in basic probability or statistics contain all the relevant mathematical expressions. We can therefore equate as many of these theoretical values to their sample counterparts as we need to determine the parameter values by solving the resulting equations. If the model has just one parameter, then it is sufficient to equate the two means and then solve the resulting equation for the parameter. If there are two parameters then we need to equate the means and also the variances, and then solve the resulting pair of simultaneous equations. Each extra parameter requires us to equate one extra pair of matching summary quantities and then to obtain the solution of progressively more equations. However, most simple models have no more than two parameters so this procedure is not generally very arduous. This method of determining parameter values is known as the *method of moments*. For example, we saw earlier that the expected value of a Poisson distribution with parameter λ is λ, so the method of moments estimator of λ for the pieces of cloth above would be given by setting λ equal to the sample mean. This is the simplest possible situation, as no solution of equations is even needed. The mean number of defects in the sample is easily found to be 1.435, so this is the method of moments estimate of λ.

The method of moments is a simple method that is implemented quite easily, but it only guarantees to match one or two summary values between sample and population. Consequently, it may lead to major discrepancies at a more detailed level, as it is entirely possible for two distributions to have very similar summary values and yet have very different shapes. This is particularly true if there is only one parameter to be estimated, as the method only matches the sample and population means and ignores all remaining characteristics. Other methods have therefore been proposed over the years, in order to take more detail into account in the matching between sample and population. One method which was popular in the early days but which has dropped out of fashion somewhat is the so-called *minimum chi-squared* method. This finds the value of λ which matches the observed frequencies and their theoretical (i.e. expected) counterparts as closely as possible, by finding the value of λ that minimises a weighted sum of squared discrepancies between the two sets of frequencies. While potentially attractive for such discrete measurement problems as the example above, there are theoretical problems with continuous measurements (caused by lack of uniqueness of intervals), so this approach has now been supplanted almost universally by the method of *maximum likelihood*. To describe this method we will first consider discrete measurements, for which the concepts are straightforward, and then turn to continuous measurements.

The *likelihood* of a sample of discrete measurements is simply the probability of the whole sample, assuming the chosen model to be the 'correct' one, and the maximum likelihood estimates of the parameters are the values that make this probability as large as possible. Let us first illustrate the procedure on the cloth defects data, and then discuss its interpretation. The first step is to calculate the likelihood, and to do this we must assume the sample to be a random one so that each observation is independent of every other one. We can then use the standard rules of probability. So for the data and Poisson model as above, each of the 92 pieces that have no defects contributes a probability $e^{-\lambda}$ to the likelihood. There are 92 such mutually independent pieces, so their joint contribution is $e^{-\lambda}$ multiplied together 92 times, which after some simplification turns out to be $e^{-92\lambda}$. Likewise, each of the 142 pieces with one defect contributes $\lambda e^{-\lambda}$, so their joint contribution is $\lambda^{142} e^{-142\lambda}$. We can similarly find the joint contributions of the 96 pieces with 2 defects, the 46 pieces with 3 defects, the 18 pieces with 4 defects and the 6 pieces with 5 defects. Finally, we again invoke the independence of pieces and obtain the likelihood by multiplying all these joint contributions together. Readers may baulk at the potential complexity of the resulting expression, but some simple algebra will reduce it to a fairly manageable form. The essential point is that it is then viewed as a function of the unknown parameter λ, and we need to look for the value of λ that makes it a maximum. This can be done either numerically using computer maximisation programs, or mathematically with the aid of differential calculus. Whichever approach we take, it turns out that for the data on defects in cloth we again estimate λ to be 1.435.

A similar approach is taken for any other set of discrete data and associated model, however many unknown parameters it contains. The likelihood will be a function of these parameters, and different sets of values of the parameters will produce different values of this likelihood. If we could write down all the possible sets of values of the parameters, and alongside each set we wrote the likelihood corresponding to the parameter values in that set, then the maximum likelihood estimates of the parameters would be the values in the set that was next to the largest likelihood. Since in general there are infinitely many such sets we cannot enumerate them in this way, so we need to use either a computer algorithm or some advanced mathematics to find the maximising set. Now, since the likelihood at a given set of values of the parameters is just the probability of the sample for those values, the method chooses the values of the parameters that ascribe greatest probability to the data that have been observed, that is, the values for which the observed data are 'most likely'. This seems to be an eminently reasonable criterion to use for determining parameter estimates—but note that it is *not* the one that produces the 'most likely' set of parameter values. Probabilities here are associated with data values, and not parameter values; this is a distinction that we will return to in Chapter 5.

When it comes to continuous measurements, there are various technical complications in any formal justification of the method so we will content ourselves with a looser and more intuitive one. Suppose that we again have a (random) sample but this time of continuous measurements, and our model has probability density function $f(x)$ which includes some parameters to be estimated. To define the likelihood of the sample we would like to multiply together the probabilities of each of the observed values assuming the model to be correct, but we saw earlier that probabilities for continuous measurements can only be obtained for intervals of values and not for individual values. Let us therefore imagine that there is an interval of very small width w centred at each of the sample values, and assume that the probability of each value is equal to the probability that it belongs to the given interval. Then the probabilities of the sample members are the areas under the curve $f(x)$ for each of the intervals, and if w is small enough the area at a particular value $x = v$ will be approximately equal to the area of the rectangle having width w and height equal to the value $f(v)$ of $f(x)$ at $x = v$. So the probability of the value at $x = v$ is equal to $wf(v)$. But w is a constant that is unchanged when different values are substituted for the model parameters, so as far as parameter estimation is concerned we can simply take the probability that $x = v$ as proportional to $f(v)$. Thus if we replace x in the probability density function of the chosen model in turn by each of the sample values and then multiply together all the resulting expressions we will obtain a value that is a constant multiple of the likelihood, so can be treated as if it *is* the likelihood as far as parameter estimation is concerned. The maximum likelihood estimates of the unknown parameters are thus given by the values that make this likelihood as large as possible.

To illustrate the procedure for continuous measurements, consider the following example. A common device when eliciting peoples' opinions about various issues is to present them with a number of statements relating to these issues, and then to determine the extent to which they agree with the statements. Their extent of agreement with each statement is obtained by providing them with a line of standard length (say 1 in., 10 cm or 1 foot) on a piece of card, marked 'totally agree' at one end and 'totally disagree' at the other, and asking them to place a cross at the point on the line that reflects their opinion. The distance from the 'totally disagree' end divided by the total length then measures the extent of their agreement, on the scale $0-1$. Suppose that 10 individuals are chosen at random, presented with a statement contrasting the policies of two political parties (e.g. 'The Conservatives are tougher on crime than Labour') and they produce the following distances:

0.9960, 0.3125, 0.4374, 0.7464, 0.8278, 0.9518, 0.9924,
   0.7112, 0.2228, 0.8609.

How might we model these responses?

We first need to come up with a sensible model, and then we need to estimate its parameters. Here the observations are continuous and on the scale $0-1$, so from earlier discussions we might consider a beta distribution as an appropriate model. Moreover, in the interests of simplicity, we would prefer to have a model containing just one parameter instead of two if possible, so might think of setting $b = 1$. If we do this, then the probability density function of the model would be

$$f(x) = ax^{a-1}.$$

The likelihood of the sample would then be found by successively replacing x in this density function by the 10 values above and multiplying the results together. We would then look for the value of a that maximised this likelihood, and it turns out that the appropriate value is 2.21. Thus our population model would have a probability density function $f(x) = 2.21x^{1.21}$.

In this example, and also the earlier ones involving parameter estimation, we have worked from first principles by applying the sample data directly to the given model, thereby either obtaining the likelihood or the summary values for use in the estimation process. However, given the relatively small set of available population models and the potentially extensive calculations involved, this is a wasteful procedure akin to 'rediscovering the wheel' every time a previously considered model is applied to a new set of data. A much more efficient procedure is to derive mathematical expressions for the maximum likelihood (or method of moments) estimates of the parameters of a particular model, in terms of a *hypothetical* set of n sample values typically denoted $x_1, x_2, x_3, \ldots, x_n$, and then to replace these hypothetical values by the actual ones obtained in any given situation. Such a mathematical expression is termed the *estimator* of the parameter in question, as opposed to the *estimate* which results when the hypothetical values are replaced by the given sample values. (We can view the estimator as a *recipe*, and the estimate as the outcome of this recipe for given ingredients, namely sample values.) This is an efficient procedure because the mathematics needs only to be done once, and typically leads to a fairly straightforward estimator that is then easily evaluated for any given set of sample values. For example, the estimator for the parameter a in the above beta distribution with $b = 1$ is given by $-\frac{1}{m}$, where m is the average of the natural logarithms of the sample members. This average is -0.4525 for the given sample values, which leads to the estimate 2.21 quoted above; given any other sample with the same model, we would need to conduct only this simple calculation and not the full likelihood maximisation.

Of course, we could equally well have used the method of moments to estimate a above. From standard theory for the beta distribution, the mean of the model is found to be $a/(a + 1)$. Denoting the mean of the sample by $\bar{x}$, equating this to the model mean and solving for a, the method of moments estimator is given by $\bar{x}/(1 - \bar{x})$. For the given sample we find $\bar{x} = 0.7059$, so on substituting this into the estimator we obtain the estimate 2.4. Thus here is a case where the method of moments and the method of maximum likelihood give different estimates of a population parameter. It therefore begs the question: How do we decide in a particular case which of several methods gives us the 'best' estimate? Answering such a question could be tackled at several levels: the theoretical, by determining which of the methods has the best general *properties*, or the practical, by determining which of the methods has given the estimate that best matches the data. As regards general theoretical properties, we need to study the behaviour of the estimator in various ways: (i) under repeated sampling from a population; (ii) as the sample size varies and (iii) under various transformations of the data. Such studies are inevitably highly mathematical so their description lies beyond

Table 3.3. Actual and theoretical distributions of numbers of defects on 400 pieces of cloth

Number of defects	0	1	2	3	4	5	Total
Actual distribution	92	142	96	46	18	6	400
Theoretical distribution	95	137	98	47	17	5	399

the scope of this book, but as a (very!) broad generalisation we might say that there is usually no contest: maximum likelihood has a number of very attractive statistical properties as well as a good track record of use, while the method of moments has no theoretical advantages but only simplicity of execution to recommend it. Maximum likelihood would therefore always be the preferred method *a priori*, but there are situations in which it is either not easy or even impossible to implement, and there may be awkward data sets where it does not produce good estimates. In these cases, we can always fall back on the method of moments.

But that still leaves the question: How can we decide whether a particular choice of model plus set of parameter estimates is a 'good' one? Let us return to the data on defects in cloth, where for the Poisson model both the method of moments and the method of maximum likelihood agree that the best choice of λ is given by the mean of the sample, 1.435. Replacing λ by 1.435 in the expression above we find that the expected number of pieces of cloth with r defects is $400 \times (1.435^r/r!)e^{-1.435}$. Working out this value for each r from 1 to 5 and rounding the results to the nearest integer, we obtain the comparison between the actual and theoretical distributions shown in Table 3.3. The discrepancy of 1 between totals in the two distributions arises partly through the rounding and partly because the theoretical distribution allows for the possibility of having some pieces of cloth with more than 5 defects on them. Thus, strictly, we should have extended the table by adding the category 'more than 5 defects' that has the actual distribution value 0 and theoretical distribution value 1. Overall, however, there seems to be good agreement between the numbers observed and the numbers derived from the model.

The natural question that might therefore now be asked is whether this agreement is close enough for the cloth manufacturer to be able to assume the Poisson model with confidence for use in any future decision-making. An associated question is: How large must the discrepancy be before we can say that the Poisson model is *not* appropriate, and indeed how do we measure this discrepancy in the first place? Answers to such questions must await the considerations of the next chapter.

Statistical Inference—the Frequentist Approach

Introduction

So far we have looked mainly at the ideas of probability, with their extension to the concept of a population of individuals having a probability distribution of some measurements of interest. However, we have also gone on to consider probability models for populations, the collection and summarisation of sample data from these populations, and the use of the sample data to guide us when formulating probability models. All these topics form the necessary base for one of the central topics in statistics, namely the use of sample data in making statements about populations. Such a process of arguing from the particular to the general is known as *induction*, and when it specifically relates to samples and populations it is termed *statistical inference*. Because we deal with random samples whose memberships (and hence whose measurements) are unpredictable, this process is always attended by uncertainty; the objectives of statistical theory are to quantify such uncertainty and hence to attach measures of confidence to all the statements that are made. Note that any argument from the population to the sample, such as the calculation of the probabilities of finding particular values of the measurement in the sample, is *deductive* and exact: if we have the right probability model, then we simply apply the mathematical rules of probability and arrive at a single incontrovertible answer. But life is such that we rarely have certainty. Instead we have to take decisions or judge between competing theories in the face of uncertainty, and it is in such circumstances that the ideas of statistical inference are invaluable.

The earliest work on what would now be classed as statistical inference dates from the late eighteenth and early nineteenth centuries, and is associated mainly with the mathematicians Adrien Marie Legendre, Carl Friedrich Gauss and Pierre-Simon Laplace. To a great extent their work overlapped, but to draw some distinctions we might say that the first-named concentrated on the ideas of errors of measurement and the concept of least squares estimation, which we will look at in some more detail in a later chapter, while the latter two were responsible for work leading up to the central limit theorem and the grounding of the normal distribution in statistics. Much of the statistics of the nineteenth century followed these ideas and thus concentrated on large-sample results. The breakthrough into the modern era came at the start of the twentieth century with the work for the Guinness brewery of William Gossett (pen-name 'student'), who developed the first small-sample results of practical use, and this was followed rapidly by the great theoretical developments of Ronald Fisher in the 1920s and of Egon Pearson and Jerzy Neyman in the 1930s. Between them these eminent statisticians supplied the framework of the frequentist approach to statistical inference that we have today. In this chapter, we will trace these developments and provide a summary of the main ideas.

Before we do this, however, we need to pause over the word 'frequentist' which is probably unfamiliar to most readers. The early pioneers of statistical inference were acutely aware that sampling is a haphazard process, and that the actual measurements provided in a given sample are governed by chance. So if we are to make statements about a population based on a sample, how are we to interpret measures of precision regarding these statements? The solution they came up with was to relate any measures of precision of a statement to the (relative) frequency with which this statement would be true *over a large number of repeated samples from the same population*. Hence the later term 'frequentist' for someone who adhered to this philosophy of interpretation. Needless to say, the term was coined when there arose competing philosophies, and we shall survey some of these in the next chapter. But for the present we fix on the frequentist approach, so we need to start by examining the implications of sampling, and to define the idea of a sampling distribution.

Sampling and sampling distributions

To take a practical example, let us suppose that an administrator at a large university wishes to investigate the distribution of distances between their home and the university of all first-year students. In principle, the administrator could extract the registration details of all the students and calculate the distance between their home and the university using appropriate software, but this might be an extremely time-consuming activity. So instead she decides to spend an hour each day for six weeks visiting one of the 30 Departments, attending a lecture for first-year students and asking a sample of students to tell her the distance between their home and the university (she is willing to accept answers to within a tolerance of a few miles). Let us suppose that there are 10,000 first-year students at the university, and the administrator's primary interest is in determining the mean and the standard deviation of the population of 10,000 distances. Assuming she can get a sample of around 30 students at each visit, she should end up with a total sample of about 1000, that is, 1 in 10 of the population. She therefore hopes that the mean and standard deviation of the sample should give good approximations to those of the population. Putting aside any questions about 'appropriateness' and 'goodness' for the time being, however, let us first examine some features of this process.

The first thing the administrator will notice on moving from Department to Department is that the set of distances collected will vary from day to day, both in their range and in their make-up. If she draws up a frequency distribution each day, then the features of these distributions will vary, sometimes quite dramatically. Assuming that there is no systematic or causal link between subjects studied and distances travelled by the students (which seems to be a perfectly reasonable assumption), the samples on different days are in effect random samples from the population of distances and the differences observed between days are simply a consequence of random variation between samples—a manifestation of *sampling variability*. There is an immediate consequence for the quantities that the administrator is interested in, namely the mean and the standard deviation. If she were to compute these two quantities for each sample, instead of waiting till she had all 1000 values, then since the samples vary in their make-up from day to day the values of the mean and standard deviation will also vary from day to day. Indeed, we could draw up a frequency distribution of the 30 means, and this would represent the *sampling distribution* of the mean. A frequency distribution of the standard deviations would likewise represent the sampling distribution of the standard deviation. In fact, we would have such a sampling distribution associated with *any* quantity calculable from the sample values, such as the proportion of sample members travelling more than 100 miles, or the maximum distance in the sample, or the sample median value.

To illustrate these and later ideas, consider the results of some computer simulations in which the distances travelled by all students in this university constitute a population having a uniform discrete distribution in the range 1–200 miles, so that the distance travelled is an integer number of miles and each value between 1 and 200 miles is equally probable. Standard theory tells us that this probability distribution has a mean μ of 100.5 miles and a standard deviation σ of 57.74 miles, so these are the characteristics of the population that are of interest. Computer routines are available on most standard software packages for drawing a sample of any size from a given probability distribution, using pseudo-random numbers as mentioned in Chapter 2, and such routines were used to draw a sample of 1000 distances from the above population. The sample mean turned out to be 101.26 and the sample standard deviation 59.39, both tolerably close to their population counterparts. Figure 4.1 shows a histogram of the relative frequencies of the different values in the sample; these relative frequencies reflect the equal values of the corresponding population probabilities, with departures from a 'flat' histogram indicating the extent of sampling variability.

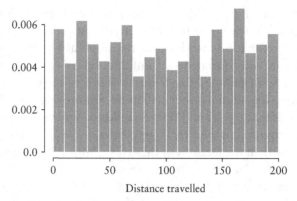

Figure 4.1. Histogram of relative frequencies for a sample from a discrete uniform distribution on the range 1–200.

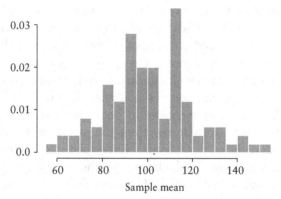

Figure 4.2. Sampling distribution of the means of 100 samples of size 10 from a discrete uniform distribution on the range 1–200.

Next, these 1000 values were divided into 100 batches of 10 values each, to mimic the drawing of 100 independent samples of size 10 from the population, and the means and standard deviations of each of these 100 samples were calculated. Histograms of the relative frequencies of these values thus provide graphical estimates of their sampling distributions; Figure 4.2 shows the sampling distribution of the 100 means, and Figure 4.3 shows the sampling distribution of the 100 standard deviations. The sample means range from just below 60 to just above 150—quite a large amount of variation—with greatest frequency between 80 and 120. The sample standard deviations exhibit somewhat greater irregularity, but also show greatest frequency in the middle of the range.

To move briefly to terminology, any quantity such as those above that is obtainable simply from sample values is termed a *statistic*. There is a sampling distribution associated with any statistic, but for it to be well

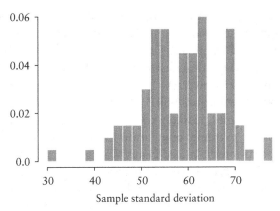

Figure 4.3. Sampling distribution of the standard deviations of 100 samples of size 10 from a discrete uniform distribution on the range 1 – 200.

defined we need to state the sample size and the population from which it is drawn. Strictly, the sampling distribution is the distribution of values of the statistic from *all possible* samples of the given size that can be drawn from the population. So if we want to study a sampling distribution we are faced with the same problem as we have with any population: we cannot ever obtain all possible samples, so we have to make do with either collecting a finite number of samples (as in the distances collected by the administrator) or formulating a probability model for the full distribution and deriving the sampling distribution theoretically or via simulation (as in our illustrations above). In fact, however, we are rarely interested in studying the sampling distribution of a given statistic per se, but usually need it as a stepping-stone to statistical inference. So we hardly ever need to collect multiple samples; the usual process is to make do with just a single sample, and attempt to model the sampling distributions of the relevant statistics. Fortunately, as we see next, there are certain results that enable us to obtain very accurate models for these distributions in certain circumstances. Moreover, the assumptions we make about the population from which we are sampling will often lead us to appropriate theoretical models for more complicated situations.

So let us continue with our general investigation of sampling distributions. Now suppose that there are three administrators, each visiting a Department each day to collect data on distances as above, but each one collecting samples of different size from the others. Suppose administrator A collects a sample of 10 distances from each Department, while administrators B and C collect samples of 20 and 100 distances, respectively. At the end of the month they compare their frequency distributions; how might we expect these distributions to differ?

Consider first the frequency (i.e. sampling) distributions of the sample means. Since the samples are all the same size for a particular administrator, the mean of the means is in effect the mean of the total collection of that administrator's individual values (this can be verified easily by writing out a few samples of simple numbers if it does not appear self-evident). So the 'centres' (i.e. means) of the three sampling distributions should be in reasonably close agreement. However, the 'widths' (i.e. the standard deviations) of the three sampling distributions will be distinctly different. Administrator A, with the smallest samples, is likely to have the biggest standard deviation. This is because a few 'extreme' population values in a small sample will have little chance to be counterbalanced in any way, so will tend to dominate and produce an 'extreme' mean. However, extreme population values in a bigger sample will be counterbalanced either by extreme values in the opposite direction or by a large excess of 'average' values, so will tend to produce an 'average' mean. Thus the bigger the samples, the more the sampling distribution of the mean will bunch in the middle and the thinner will be its 'tails'. So administrators B and C will have progressively smaller standard deviations of their sampling distributions. In fact, theory shows that if μ and σ are the mean and standard deviation, respectively, of the population, then the mean and standard deviation of the sampling distribution of the sample mean will be μ and $\sigma/\sqrt{n}$. Moreover, theory also tells us that if the original population has a normal distribution, then the sampling distribution of the sample mean also has a normal distribution. Allied to the above two values, this describes its shape exactly.

But there is an even more remarkable theoretical result for the sampling distribution of the sample mean, and this is the famous central limit theorem which says that the sampling distribution of the sample mean becomes closer and closer to a normal distribution as the sample size increases, *whatever the nature of the original population*. This means that we can assume a normal model for the sampling distribution of the sample means, provided that the sample size is 'big enough'. How big is 'big enough'? This depends on the nature of the population from which we are sampling. If the population is reasonably symmetric and does not have any unexpected features, then a sample as small as 10 or 12 might be big enough. In general though, if we know very little about the population, then we should aim for samples of at least about 25–30 members if the central limit theorem is to be usable.

To illustrate these ideas, let us return to our uniform discrete distribution (definitely not normal!) of distances travelled. The previous sampling distributions mimicked those expected for administrator A, so to mimic administrators B and C two further samples, of sizes 2000 and 10,000, were drawn from this population. The former was divided into 100 batches each of size 20, and the latter into 100 batches each of size 100. Let us just focus on the sampling distributions of the means in each case. The theory given above, linked to the population values $\mu = 100.5$ miles and $\sigma = 57.74$ given earlier, suggests that the three sampling distributions should all have mean $\mu = 100.5$, with standard deviations of $57.74/\sqrt{10} = 18.26$, $57.74/\sqrt{20} = 12.91$ and $57.74/\sqrt{100} = 5.77$, respectively. In fact, the three means turned out to be 101.26, 101.05 and 100.51 (drawing progressively closer to 100.5 as sample size increases) while the three standard deviations were 19.18, 12.72 and 5.69 (all close to their population values, and getting closer as sample size increases). The sampling distribution of the means has already been shown in Figure 4.2 for samples of size 10, so Figures 4.4 and 4.5 show the corresponding distributions for samples of size 20 and 100, respectively. The reduction in spread of values and approach to normality of shape as sample size increases is very evident.

But the usefulness of the central limit theorem does not stop there, as the normal approximation holds good for any type of mean, whether it is a weighted or unweighted mean, and moreover it extends to any statistic that can be expressed as some form of mean. Thus, for example, suppose

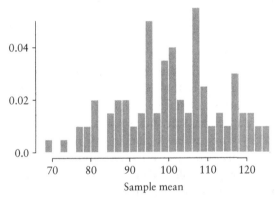

Figure 4.4. Sampling distribution of the means of 100 samples of size 20 from a discrete uniform distribution on the range $1-200$.

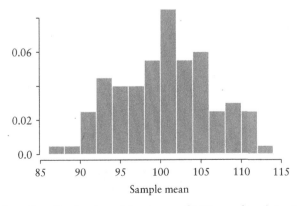

Figure 4.5. Sampling distribution of the means of 100 samples of size 100 from a discrete uniform distribution on the range $1-200$.

that the administrator is interested in the proportion of students who travel more than 100 miles to the university. Let us define a measurement, say X, as having the value 1 for a particular student if that student travels more than 100 miles to the university and the value 0 otherwise. Then the average of X in the population is the proportion of students in the population who travel more than 100 miles, and the average of X in the sample is the corresponding sample proportion. So the central limit theorem applies to sampling distributions of proportions as well, and we can assume a normal model once the sample is bigger than about 25. To illustrate this, let us return for a last time to the theoretical population of distances travelled, where for the uniform distribution the probability of obtaining a value greater than 100 is 0.5. One hundred samples of size 50 were drawn, and the proportion of values greater than 100 was calculated for each sample. Figure 4.6 demonstrates the approximate normality of these values, the correspondence with a bell-shaped normal curve being marred only by the reversed frequencies in a couple of intervals just below 0.5. Note also the spread of values around the expected value of 0.5, the bulk of which fall between 0.4 and 0.6.

The central limit theorem therefore opens the door to a large number of results in standard frequentist statistical inference. Unfortunately, however, we cannot in general appeal to the central limit theorem to help us in modelling sampling distributions of statistics that are not expressible as means of sample values. In particular, statistics such as the sample variance or the sample standard deviation are awkward in this respect. Although the same general principles apply with regard to the width of the sampling distributions (i.e. the greater the sample, the 'narrower' the sampling distribution), we cannot say anything about the overall shape of the distributions unless the population has a normal distribution itself. Even then the sampling distributions differ from normality, so we need to refer to specific theoretical results.

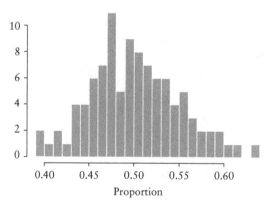

Figure 4.6.   Sampling distribution of the proportions of values greater than 100 in 100 samples of size 50 from a discrete uniform distribution on the range $1-200$.

A final point about sampling distributions concerns terminology, as it may have occurred to the reader that there is potential for considerable confusion with some of the quantities we have been discussing. We start with a sample from the population, so we can compute the mean and standard deviation of the sample values. But the sample mean and standard deviation are both statistics so both have sampling distributions, which describe their likely variability from sample to sample. These sampling distributions each have their own means and standard deviations—so when we talk about a mean or a standard deviation, we must be very clear whether we are referring to the original sample or to one of the sampling distributions. To help with this, and to avoid having to describe the quantity in tortuous detail, the standard deviation of the sampling distribution of a statistic is termed the *standard error* of that statistic (because the standard deviation of the sampling distribution quantifies the likely variability, or error, in that statistic). So when we talk about the standard error of the sample mean, we mean the standard deviation of the sampling distribution of the sample mean. Similarly for any other statistic, such as the standard error of the sample standard deviation, the standard error of a sample proportion, and so on.

Statistical inference

We have previously said that the process of statistical inference is to do with making statements about a population from information about it contained in a sample. So we now need to consider what types of statement we might wish to make, and how to convert these statements into a form that is amenable to statistical treatment. Essentially there are two distinct situations: if we do not know anything about a population, or more precisely some specific feature of the population, then we need find out about it from the sample; alternatively, if we hold some particular view, or have some specific theory, about the population then we need to see if the sample data uphold this view/theory or not. The first situation leads us to the process of *estimation*, the second one to the process of *hypothesis testing*. We give a few general points about each of these areas here, and then consider them in some more detail below.

The main point about both areas is that in order to turn them into a form that is amenable to a statistical approach, we must somehow express the statements in terms of the probability model chosen for the population. Since these models are usually well defined apart from lack of precise knowledge of some or all parameters, the statements will inevitably focus on these parameters. In the first situation we will want to estimate the parameters from the sample data, and in the second case we will need to express our theory in terms of the parameters before we can see whether the sample data support it or not.

On the face of it, the estimation problem seems straightforward. For example, the manufacturer of camshafts may be prepared to assume a normal probability model for the population of camshaft lengths, so all he needs to find in order to specify the population exactly is the mean and standard deviation of the population. We have noted earlier that there are various general estimation methods (such as maximum likelihood, method of moments and so on) that we can employ in order to estimate parameters from data, and for the normal model these methods all tell us that the best estimates are the mean and standard deviation of the sample. So is the problem solved? Well, not really, because all we have is a single 'guess' at the value of each parameter. Admittedly, this is an educated guess because it employs the sample data along with a 'good' method of estimation, but there is negligible chance that we have actually hit on the 'correct' values. Why is this? Well, the preceding section has shown that if we were to take another sample we would have obtained two *different* estimates, and another two different estimates with an yet further sample, and so on. In general, many different samples will yield many different estimates and therefore any single one has a negligible chance of being correct. So the implication is that quoting a single guess at each parameter value is not good enough—if we do not have much confidence that the quoted value is correct then we must either give some indication of its likely error, or instead quote a *range* of values within which we are reasonably sure that the correct value *does* lie. This is the subject matter of *confidence intervals*, which we discuss below.

The hypothesis testing scenario is perhaps straightforward to state, but working out a strategy to conduct the test is maybe less straightforward. We want to know whether the sample data support some preconceived theory we hold about the parameters of the population model. For example, the camshaft manufacturer may be working to some international standards and hence believes that the population of camshaft lengths has some specified mean value. How does he decide whether the sample data support this belief or not? One possible way forward might be to ask the question: how likely are the values observed in the sample to occur if the belief is correct? If we can find a numerical value of the relevant probability, then we might be able to assess whether the belief is supported or not: a high probability of obtaining the data supports the belief, a low one does not. But what is 'high' and what is 'low'? Is it in fact possible to obtain the probability of the data in the light of the belief? Will the probability depend on possible competitors to this belief? These are the sorts of questions that we consider in the *hypothesis testing* section below.

Confidence intervals

We have seen above that finding an estimated value of some population parameter θ, say, from sample data is not a very satisfactory procedure because we can be (almost) sure that our estimate is not correct, and it has been suggested that a much more preferable procedure is to find a range of values within which we have high confidence that the true value of θ actually lies. The set of values ranging from a to b can be conveniently denoted (a, b)—so that, for example, $(1.76, 24.2)$ contains all values between 1.76 and 24.2—and if the true value of θ lies somewhere in this range then mathematically we write either $\theta \in (a, b)$, where '$\in$' denotes 'belongs to' in the terminology of sets of values, or $a < \theta < b$, which denotes that θ is greater than a but less than b. Now how can we express the fact that we have 'high confidence' that θ lies somewhere in this range? Let us for a moment forget the technicalities of probability as discussed earlier, and revert to a purely intuitive approach. In many facets of everyday life we make statements such as that we are 'almost sure' or 'reasonably confident' of something, and what we are doing implicitly is quantifying our degree of belief about the relevant event A. So by 'almost sure' we mean that we think the probability of A is very close to 1—say 0.99 or 0.95—whereas by 'reasonably confident' we mean that the probability of A is somewhat lower but still much greater than 0.5—say 0.9 or 0.8. Consequently, if we were to say that we had high confidence that the true value of θ lay between a and b, then mentally we would probably interpret this as saying that $P\{\theta \in (a, b)\}$ had some value c close to 1.

There are three values of c that have gained a traditional place in such calculations, namely 0.99, 0.95 and 0.9, and these are usually expressed as percentage confidence—namely 99% confident, 95% confident and 90% confident. Informally, they might be considered to correspond to the statements 'highly confident', 'confident' and 'somewhat confident', respectively. As we shall see later in this chapter, they also link directly to their complements 1%, 5% and 10%, traditionally adopted as significance levels in hypothesis testing. These values are of course entirely arbitrary and could have been supplanted by any others, so we might ask why they have assumed such iconic status. As with many such 'traditions', various suggestions have been put forward but the true reason is now lost in the mists of time. A possibility is that it dates from a copyright problem arising between those two archrivals of the early twentieth century, Karl Pearson and Ronald Fisher. The former was the first to publish comprehensive statistical tables ('*Tables for Statisticians and Biometricians*' in 1914), and these tables encompassed an extensive set of significance levels. Fisher's '*Statistical Methods for Research Workers*', published in 1925, and '*Statistical Tables for Biological, Agricultural, and Medical Research*' published with Frank Yates in 1938, included much more compact tables focusing on just the above few significance levels and in particular giving special status to the 5% level. It is thought that Fisher deliberately chose only a few such values in order to avoid a copyright wrangle with Pearson. However, his influence was so massive throughout the 1930s, 1940s and 1950s that his approach prevailed and these few values became firmly accepted as the norm.

So returning to the main theme, a sensible objective in estimation problems is to find the appropriate values of a and b that correspond to the chosen confidence level c for any given population parameter. This is not the place for a detailed technical development of the theory, but perhaps an outline sketch of one general approach plus a couple of simple examples would be useful. The actual details differ from model to model and parameter to parameter, but the essential first step in this method is to find what is called a *pivotal quantity* for the relevant parameter/model combination. A pivotal quantity is a function of the parameter in question, sample statistics and constants that has a fully determined sampling distribution, so that if the chosen confidence level is c we can find exactly (often with the help of statistical tables) the end points of an interval within which the pivotal quantity lies with probability c. Algebraic manipulation of the resulting probability statement will then usually yield the required values a and b.

Consider the camshaft lengths again. If we are prepared to assume that the population of camshaft lengths is normal, with some mean μ and some standard deviation σ, then we have seen above that the sampling distribution of the sample mean is also normal, with mean μ and standard deviation $\sigma/\sqrt{n}$. Even if we do not make the normality assumption for the population, the central limit theorem ensures that with a sample size as large as 100, the sampling distribution of the mean should be very close to a normal distribution. Hence, if we denote the sample mean by $\bar{x}$, the quantity $z = (\bar{x} - \mu)/(\sigma/\sqrt{n})$ has a standard normal distribution. If the value of the population standard deviation σ is known (or if we assume it to be known), then the quantity z is a pivotal quantity for μ as defined above, because all the quantities in it apart from μ are either known constants (σ, n) or obtainable from the sample $(\bar{x})$ and the distribution of z is fully determined with probabilities obtainable directly from tables of the standard normal distribution. If we choose a confidence level of 95%, that is, $c = 0.95$, then we find from these tables that $P(-1.96 < z < 1.96) = 0.95$. Replacing z inside this probability statement by the function above, and carrying out a bit of algebraic manipulation, shows that this statement is exactly equivalent to the statement

$$P\left\{\bar{x} - 1.96\frac{\sigma}{\sqrt{n}} < \mu < \bar{x} + 1.96\frac{\sigma}{\sqrt{n}}\right\} = 0.95.$$

Thus by comparison with the definition above, the end points of the 95% confidence interval for μ are $a = \bar{x} - 1.96(\sigma/\sqrt{n})$ and $b = \bar{x} + 1.96(\sigma/\sqrt{n})$. For the camshaft data we had a sample size $n = 100$, and we found that the mean was $\bar{x} = 600.142$. If the population standard deviation is $\sigma = 1.5$, then substituting all these values into the expressions above gives $a = 599.848$ and $b = 600.436$. These are the end points of our 95% confidence interval, so according to our informal categorisation above, we are 'confident' that the population mean μ of camshaft lengths lies between 599.848 and 600.436. If we want to be 'highly confident' we need to increase our confidence level to 99%, which requires replacement of 1.96 in these calculations by 2.5758. Doing this changes the confidence interval to (599.756, 600.528), which is slightly wider than before. Conversely, if we reduce our confidence level to 90% then we replace 1.96 by 1.645, and we are then 'somewhat confident' that μ lies between 599.895 and 600.389—a narrower interval than before.

These calculations illustrate a general feature of confidence intervals, namely that for a fixed sample size n raising our confidence level will *increase* the width of the interval while lowering our confidence level will *decrease* the width. This accords with intuition: the more values in the interval, the more confident are we that the interval contains the true value of the parameter. It is instructive to take this argument to its extremes. If we want to be *certain* of including the parameter we need to raise our confidence level to 100%, which requires replacing 1.96 by a very large number and this in turn makes the interval so wide as to be useless. On the other hand, if we reduce our confidence level to near zero, then we replace 1.96 effectively by zero and our 'interval' shrinks to the single point $\bar{x}$. This supports our previous assertion that we have virtually no confidence that $\bar{x}$ is the true population mean. Note also that there is another way of reducing the width of the interval rather than changing our confidence level, and that is by increasing the sample size n. The form of the interval above shows that its width is governed by a factor that includes the reciprocal of n. So as n increases (while the confidence level and σ remain unchanged), this factor becomes smaller and hence the interval becomes narrower. The extreme situation again accords with intuition: as n approaches infinity (i.e. as the sample approaches the population), the sample mean should approach the population mean in value.

These general features can be illustrated graphically using the values $\bar{x} = 600.142$ and $\sigma = 1.5$ for the camshaft data as the basis of confidence intervals for the population mean. First suppose that we consider the sample size to be fixed at $n = 100$. The limits of the confidence interval for any chosen confidence level between 0% and 100% can then be read off from Figure 4.7 as the x values of the two curves at that confidence level. The rapid widening of the intervals as the confidence level approaches 100% is very evident. Next suppose that we fix the confidence level at 95%, but let the sample size vary from 1 to 100 (while keeping the values of $\bar{x}$ and σ fixed). Figure 4.8 shows the limits of the confidence interval at a given sample size as the x values of the two curves at that value of n. This time the width of the interval decreases rapidly for small sample sizes, but then stabilises and only shows very gradual decrease beyond about $n = 50$. Considering effects of sample size on width of intervals in this way can be worthwhile in situations where costs of sampling are critical.

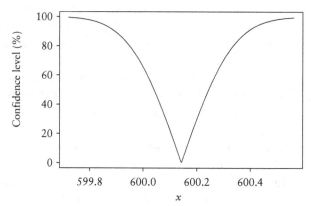

Figure 4.7. Confidence limits for the population mean of camshaft lengths, for confidence levels ranging from 0% to 100%.

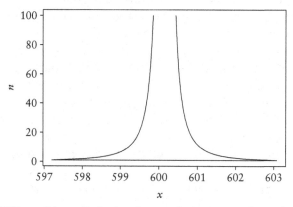

Figure 4.8. 95% confidence limits for the population mean of camshaft lengths, for sample sizes ranging from 1 to 100.

However, the reader may be worried about one aspect of the above calculations, namely that we have assumed the population standard deviation to be known to equal 1.5. What happens if we do not know this value (which is the most likely scenario in practice)? The sample standard deviation was found to be 1.34—can we use this value instead of 1.5 in our calculations? Well, clearly, if we do not know the value of σ the next best thing is to use the sample standard deviation in its place. But using a sample value in place of a population parameter *introduces extra uncertainty*, for the simple reason that we are almost sure (see above) that this is *not* the true value—a different sample will yield a different value. We may be able to get away with it if the sample is large, as then (see above again) any confidence interval for the population standard deviation should be narrow and the sample estimate should be very close to the population value. So for the sample of 100 camshaft lengths we will not go wrong too far—but what could we do if the sample were rather smaller, say just 20 values?

This was the point at which William Gossett made his great contribution that brought statistics into the twentieth century. Previously all confidence interval construction had been based on large samples, normality and the central limit theorem. Gossett's breakthrough was to realise that if σ is replaced by the sample standard deviation, say s, in the calculation of z above, then the resulting quantity no longer has a standard normal distribution. Instead, it has a distribution that allows larger values to occur, reflecting the greater uncertainty caused by estimation of σ. However, this uncertainty will decrease as the sample size increases, and eventually it should approach the standard normal distribution for large enough samples. Gossett proceeded to derive this distribution theoretically, and to specify how relevant probabilities should be calculated. It has become known as *student's t distribution* (after his pen-name 'student'); it depends on a quantity termed the *degrees of freedom*, which is given by $n - 1$ for calculating confidence intervals for μ, and tables of this distribution can be used in much the same way as tables of the standard normal distribution for obtaining relevant probabilities. The practical effect is that values such as 1.96 above are modified, being larger in all cases but with the difference from 1.96 shrinking as n increases. For example, the 95% value is 4.30 when $n = 3$; 2.26 when n is 10; 2.09 when n is 20; 2.02 when n is 40 and 2.00 when n is 60. So by the time the sample exceeds 60 individuals, the normal confidence intervals are perfectly adequate.

Following Gossett, the early years of the twentieth century saw the derivation of many other small-sample results that enabled exact confidence intervals to be obtained for various other population parameters in common demand, such as variance and ratios of variances using distributions such as the chi-squared and F distributions.

Unfortunately, once one passes outside the set of such standard problems, existence of a pivotal quantity is not guaranteed; indeed such quantities rarely exist for more complicated situations in a form that enables a confidence interval to be found by simple algebraic manipulation. In such cases, there are two possibilities open to the statistician. If the exact sampling distribution can be obtained for an estimator of the unknown parameter, then it is often possible to find an interval of given probability content from this distribution whose end points are functions of the unknown parameter. Graphing these functions and observing where they are cut by the specific value of the estimator in a given sample will yield the required confidence interval. Alternatively, appealing to the general theory of maximum likelihood estimation as developed by Ronald (later Sir Ronald) Fisher will yield approximate confidence intervals (valid for large samples) for virtually any parameter or population model. However, it is not our purpose here to delve into technicalities; most standard texts on statistical inference contain all the relevant details.

Interpretation of confidence intervals

Everyone is intuitively comfortable with a statement such as 'I am highly confident that the population mean μ lies between the two values a and b', but when dealing with scientific enquiry we need to be clear as to precisely what it means. Moreover, we started the chapter by saying that the methods of inference we would discuss were frequentist ones, namely ones that relied on the concept of repeated sampling from the same population for their interpretation. So we now need to look at confidence intervals critically, and ask how they can be interpreted in these terms.

In deriving the confidence interval for the mean μ of the population of camshaft lengths, we arrived at a probability statement of the form

$$P\left\{\bar{x} - 1.96\frac{\sigma}{\sqrt{n}} < \mu < \bar{x} + 1.96\frac{\sigma}{\sqrt{n}}\right\} = 0.95,$$

from which we deduced the end points of the 95% confidence interval for μ as $a = \bar{x} - 1.96(\sigma/\sqrt{n})$ and $b = \bar{x} + 1.96(\sigma/\sqrt{n})$. On the face of it, this is just a probability statement and so we should be able to apply a relative frequency argument in order to interpret it. However, on closer inspection we see that it is not a probability statement in the usual form of such statements. To see the difference, consider the probability distribution shown in Table 3.1 for the score B on throwing two dice. From this probability distribution we can write down a probability statement such as

$$P(4 \leq B \leq 7) = \frac{18}{36} = 0.5,$$

and we would have no trouble in interpreting it in terms of repeated sampling: In a large number of throws of the two dice, a score somewhere between 4 and 7 (inclusive) would occur half the time.

So why is there a problem with the probability statement leading to the confidence interval? The essential difference is in the nature of the inequalities inside the probability in the two cases. For the score on two dice, the *random* quantity from trial to trial is B (i.e. the score) and this occurs in the *middle* of the inequality, whereas the *fixed* values 4 and 7 occur at the two *ends* of the inequality. So the interval (4, 7) remains fixed from trial to trial, and over a large number of trials B falls in it 50% of the time. However, for the probability statement leading to the confidence interval, the positions of the random and fixed quantities are interchanged. In the centre of the inequality we now have the population mean μ. This is of course *fixed* (although it is unknown, as it is the quantity we are trying to estimate). The 'trials' in this case are the samples of size n that we could be collecting and that yield values of the sample mean $\bar{x}$. These are the values that make-up the sampling distribution of $\bar{x}$, so they and hence the *ends* of the inequality will now vary from sample to sample. Thus instead of a fixed interval and a random quantity, which falls inside or outside the interval on any trial, we now have a fixed (but unknown) quantity and a random interval which either includes or does not include the fixed quantity for any trial (i.e. sample).

Of course, the population mean μ is unknown, so we can never tell whether it is actually in the interval we have computed or not. However, the theory tells us that it will be included in 95% of the intervals calculated over a large number of samples of given size from the same population. This is why we are 95% confident that it is included in the one sample to which we are generally restricted. Corresponding interpretations apply to all confidence intervals, whatever parameters or populations we are dealing with, and the percentage inclusion rate is varied appropriately for 90% or 99% (or indeed any other) confidence levels.

The above interpretation of a confidence interval is not perhaps the most obvious one that springs to mind, so not surprisingly mistakes are frequently made when confidence intervals are interpreted. The most common mistake is to interpret the confidence interval as if it were a traditional probability statement. The effect of this is to ascribe a probability to μ on the basis of the *single* interval that has been calculated, namely to imagine that there is a distribution of potential values of μ and a 95% probability that its actual value lies in the computed interval. While this is an interpretation that many researchers would *like* to place on the interval, it is certainly incorrect within the frequentist approach where there is only ever *one* value, albeit unknown, of μ. An associated mistake when interpreting a confidence interval is to somehow 'share out' the probability across the interval. For example, suppose that a study has been made into the prices of houses in two localities A and B, and $(-16.3, 32.6)$ is a 95% confidence interval for the true average difference in prices A–B (in thousands of pounds). Since there are twice as many positive as negative values in this interval, a common temptation is to say that it is twice as likely for the average price in A to be greater than the average price in B than for the converse to be the case. This interpretation in effect ascribes a uniform probability to μ across the interval, which is inappropriate; the correct view is simply that the true average difference lies *somewhere* in this interval with 95% confidence.

If the researcher wishes to attach actual probabilities to parameters and hence intervals, then an alternative approach is necessary and Bayesian methods as described in Chapter 5 must be used.

Hypothesis testing

Basic ideas

The testing of hypotheses is perhaps the most common pursuit in scientific and social investigations, and a large proportion of articles in research journals report the results of at least one hypothesis test. The central question in hypothesis testing is deceptively simple to ask: do the data support my theory or not? Yet it is the area that has arguably produced the most involved terminology, the most tortuous procedures, and the most convoluted reasoning in a century of statistical practice. Misapplication of statistical tests can, and does, lead to misleading conclusions and potentially disastrous consequences. Yet the basic ideas are reasonably straightforward, and the structure is not complicated. So let us try here to provide some solid foundations, an appreciation of the necessary framework, and an overview of the logic of the frequentist approach. As with confidence intervals, we leave the details of specific tests for more technical expositions elsewhere.

Consider a very simple situation: I have a suspicion that a particular coin is biased when spun, and instead of having an even chance of falling either 'heads' or 'tails' it is more likely to fall 'heads'. How do I check my theory? The obvious thing to do is to spin the coin, and see how many times it comes down 'heads' and how many times it comes down 'tails': the more 'heads' there are, the more evidence there is for my theory. Moreover, the more times I spin it, the more evidence I can gather, so it is clearly in my interest to spin it a large number of times. Likewise, in serious research, the more data that can be collected, the more evidence for or against theories can be gathered and the more certainty can we have in the results. But resources, time, enthusiasm, obstacles, and life, in general, usually conspire to leave us with less evidence than we would like. So let us suppose that some external event (independent of the results obtained to that point) causes me to stop spinning the coin after, say, 12 spins, at which point 9 of my 12 spins have resulted in a 'head'. Does this support my theory that the coin is biased in favour of 'heads'?

As with confidence intervals, we immediately hit a conflict between the sort of inference we would *like* to make and the sort that the frequentist approach permits us to make. Intuitively, we would like to make a statement about the probability that our theory is correct, given the observed data. Unfortunately, under the repeated sampling view, we can only make probability statements about data given models or theories, and not about theories given data. In the confidence interval situation we have just a single (unknown) parameter value and not a distribution of possible parameter values, and likewise here we just have a single theory and not a distribution of theories that we can select from. So given that we must focus on the data, the natural approach would be to ask what the chance is of getting the observed result if my theory is correct, and accept the theory if this chance is high. But here we hit another snag, because to find a numerical value for any probability we need exact mathematical specifications. Unfortunately, my theory only goes as far as saying that the coin is biased—it does not specify *how* biased, so the specification is not complete. Clearly, the probability of getting 9 'heads' out of 12 spins will increase as the bias, namely the probability of getting a 'head' in one spin, increases. Without knowledge of this latter probability I am sunk, and cannot compute the probability I need. A little thought will convince the reader that there is only one situation in which I *can* compute the probability of obtaining 9 'heads' in 12 spins, and that is when the probability of a 'head' in one spin is exactly the same as the probability of a 'tail'—namely when the coin is *unbiased*, or the opposite of my theory. So we are led to the first general tenet of hypothesis testing within the frequentist approach, namely that we usually have to find the probability of obtaining observed results if our theory is *not* true, and take a *low* probability as evidence in favour of our theory. The spinning of a coin is an example of a Bernoulli trial as defined in the previous chapter, so the probability of obtaining a specified number of 'heads' is given by the binomial distribution. For 12 spins with an unbiased coin the parameters of the distribution are $n = 12$ and $\theta = 0.5$, so we easily find the probability of 9 'heads' to be 0.054.

So is this the end of the story, and all that remains is to decide how small the probability has to be in order for us to conclude that the data support our theory? Unfortunately, no—but to see why not it is helpful to consider another example, this time focusing on continuous measurements. Let us revisit the camshaft lengths again, and suppose that a government quality directive requires the average length of all camshafts to be greater than 600mm. The manufacturer believes that he is meeting this requirement, so in statistical terms his theory is that the population mean μ is greater than 600.00. Do the sample data given earlier support this theory or not? The *prima facie* evidence is favourable, as the sample mean is 600.14, but is this evidence sufficiently strong?

Following the same reasoning as above, the manufacturer's theory can be satisfied by many different values of μ and we do not know which one to adopt when trying to compute probabilities assuming the theory to be true. So we need to assume that the theory is not true. But here we find exactly the same problem: if the theory is not true then we can still have many different values of μ (anything less than or equal to 600.00) and we are equally unable to calculate any probabilities. To be able to do so, we need a *unique* value of μ at which to fix the mean of the population for testing the theory. The answer is to use the value of μ that is *nearest* to the ones for which the theory is true, on the grounds that if we can show the theory to be supported for this value, then it is certain to be supported for the other possible values. In the camshaft example this value is 600.00—if the sample data have low probability of occurrence when $\mu = 600.00$, then they will have even lower probability of occurrence when μ is less than 600.00.

So we appear to have overcome the obstacles, and we can follow through with the previously outlined procedure in order to test the manufacturer's theory. We need a single summary statistic from the data that will give us a mechanism for conducting the test, and that statistic is clearly the sample mean $\bar{x}$: the greater the excess of $\bar{x}$ over 600.00, the more support there is for the theory. So to follow through the previous steps of the hypothesis testing argument, we need to use the sampling distribution of $\bar{x}$ to work out the probability that the sample mean $\bar{x}$ equals 600.142 if the population mean μ equals 600.00. A low value of this probability will give support to the manufacturer's theory. For the moment, let us assume as we did in the confidence interval section that the standard deviation σ of the population is known to be 1.5. In that section we also saw that the quantity $z = (\bar{x} - \mu)/(\sigma/\sqrt{n})$ has a standard normal distribution, so we can use this result to try and calculate the required probability. We substitute the values $\bar{x} = 600.142, \mu = 600.00, \sigma = 1.5, n = 100$ in this expression, and find that $z = 0.947$. So we need to find the probability of this outcome for the standard normal distribution.

But this is where we hit our second major problem. Recollect from the previous chapter that any population model for a continuous measurement involves a probability density curve, and the probability that the measurement lies in a specified interval is given by the area of this interval under the curve. So for a continuous measurement, the probability that it equals a specified value (such as 0.947) is always zero. We need always to define an appropriate *interval* for a probability to be computable with a continuous measurement. What would be an appropriate interval in our example above? The answer comes when we examine carefully what we are looking for. We assume that our theory is not true, and choose a summary statistic that will provide evidence in favour of our theory when it takes values that are either untypical or extreme if the theory is not true. So what we are actually asking when we conduct the test is: if the theory is not true, what is the probability of obtaining a value *at least as extreme* as the one we have observed. This is because there would be stronger evidence in favour of our theory if more extreme values were observed. In the camshaft example, values of the sample mean greater than 600.142 would provide greater support for the manufacturer' theory, so what we need from tables of the standard normal distribution is the probability that z is *greater than or equal to* 0.947. This turns out to be 0.172, or 17.2%. In other words, about 17 out of every 100 samples of size $n = 100$ would provide a sample mean $\bar{x} \geq 600.142$ when the manufacturer's theory was *not* true.

The greater the value of the sample mean, the smaller will be the probability of exceeding it if the population mean μ equals 600.0. In fact, it is easy to obtain the relevant probability from normal tables for any value of the sample mean, and Figure 4.9 shows the rate at which this probability decreases for sample mean values from 600.0 to 601.0 if the values of σ and n remain unchanged.

We consider below how this probability value can be interpreted, but first note that the reasoning for continuous measurements needs to be carried over

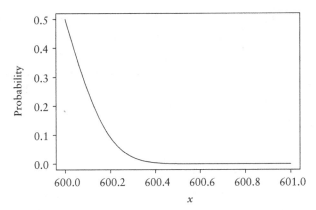

Figure 4.9. Probability of exceeding the sample mean if the population mean equals 600.0, for camshaft length parameters and sample mean values between 600.0 and 601.0.

to discrete measurements also if there is to be a general theory of hypothesis testing. So with the earlier coin spinning example, we noted that 9 out of 12 spins landed 'heads' and that the more heads there were, the greater the evidence that the coin was biased. Instead of calculating the probability of obtaining exactly 9 'heads' with an unbiased coin, therefore, the evidence should be given by the probability of obtaining 9 *or more* 'heads' (i.e. 9, 10, 11 or 12 'heads'). The binomial distribution is used for each of these outcomes, giving a revised probability of $0.054 + 0.016 + 0.003 + 0.000 = 0.073$.

Formal notation; significance level

Before considering the extent to which probabilities such as 0.073 or 0.172 provide or do not provide evidence for a theory, let us formalise some of the above discussion and express it in the standard terminology of hypothesis testing. Since we are dealing with *statistical* tests of hypothesis, every theory we wish to test has to be converted into one or more values of a parameter of a population model. It has been stressed that to seek evidence in favour of a particular theory, we usually have to calculate probabilities of sample values assuming that the theory is *not* correct. So we generally have to contrast two possible sets of values of the population parameters: values if the theory is *not* correct, and values if the theory *is* correct. The former values constitute what is usually called the *null* hypothesis, denoted by H_0, while the latter comprise the *alternative* hypothesis, denoted by H_1 or H_a. These terms derive from the belief that the null hypothesis usually represents the 'neutral', 'current', or perhaps 'undesirable' state of the system, which will remain in place until evidence is found to enable it to be supplanted by the state contained in alternative hypothesis. In our two examples above, we have $H_0: \theta = 0.5$ versus $H_1: \theta > 0.5$ for the coin spinning and $H_0: \mu = 600.00$ versus $H_1: \mu > 600.00$ for the camshaft lengths.

The evidence is given by the probability of obtaining a value of the chosen statistic that is as extreme or *more extreme in the direction of the alternative hypothesis* than the value obtained in the sample, assuming that H_0 is true. For the coin spinning example this is the probability that the number of 'heads' in 12 spins is greater than or equal to 9 when the coin is unbiased, while for the camshaft lengths it is the probability that the sample mean is greater than or equal to 600.142 when μ is 600.00. This probability is termed the *significance level* of the test. When is this evidence sufficiently persuasive? Perhaps the camshaft lengths give us something of a clue, because the reader might have noted that we have used the same quantity, $z = (\bar{x} - \mu)/(\sigma/\sqrt{n})$, to calculate this evidence that we did to obtain a confidence interval for the population mean μ. So there is some connection between the 'probabilities' employed in confidence intervals and those in tests. In fact it is an inverse relationship, because for confidence intervals we seek large probabilities, but for tests we look for small probabilities as evidence. Thus, traditionally, the values 0.1, 0.05 and 0.01 are used to denote 'some evidence', 'enough evidence' and 'very convincing evidence' in favour of the alternative hypothesis, as the complements of 0.9, 0.95 and 0.99 in confidence interval calculations. In our examples above, we have some evidence in favour of the theory that the coin is biased but not enough evidence in favour of the manufacturer's claim regarding camshaft lengths. Often, the evidence is turned into a course of action by taking the asymmetric nature of the two competing hypotheses into consideration. If the significance level drops below one of these fixed values, we *reject* the null hypothesis in favour of the alternative at the percentage equivalent of that significance level, but if the significance level is greater than the fixed value we *do not reject* the null hypothesis. The implication in the latter statement is that we will continue to believe the null hypothesis value until we obtain sufficient evidence to reject it. So we would reject the null hypothesis of unbiasedness of coin in favour of the alternative hypothesis at the 10% level of significance but not at either the 5% or the 1% levels, while we would not reject the null hypothesis for the camshaft lengths at any of the three significance levels.

One-tail and two-tail tests

Let us return to the coin spinning example. Previously it was supposed that I have a suspicion that the coin is biased in favour of 'heads'. Now suppose that I do not have as much prior belief, but from handling the coin and testing its weight I simply have a feeling that it is biased without knowing in which direction. If I want to test this theory my null hypothesis (theory not true) remains $H_0: \theta = 0.5$, but the alternative (biased but in an unknown direction) now becomes $H_1: \theta \neq 0.5$. Clearly, spinning the coin 12 times and noting the number of 'heads' is still a valid statistic but now evidence against the null hypothesis will come from *either* a lot of 'heads' *or* a lot of 'tails' (i.e. very few 'heads'). Since we do not know *before* we spin the coin which of these two possibilities will occur, we have to allow for both of them in our calculation of the evidence. Suppose we now spin the coin 12 times and again obtain 9 'heads'. On the face of it this is the same as before, but the evidence has changed because the alternative hypothesis has changed. To quote the relevant sentence from above, the evidence is given by the probability of obtaining a value of the chosen statistic that is *as extreme or more extreme in the direction of the alternative hypothesis* than the value obtained in the sample. With 9 'heads', a result as extreme or more extreme in the direction of the alternative hypothesis would now be *either* 9 or more 'heads' *or* 9 or more 'tails', because 9 'tails' is as extreme as 9 'heads' when we do not know the direction of bias. But 9 or more 'tails' is the same as 3 or fewer 'heads', so we want the probability that the number of 'heads' is one of 0, 1, 2, 3, 9, 10, 11 or 12. This requires probabilities from both ends (or tails) of the binomial distribution, so the test is known as a two-tail test, by contrast with the previous one-tail test when we just wanted the probability of 9, 10, 11 or 12 'heads'. It turns out that the binomial distribution for $\theta = 0.5$ is symmetric, so the probability of 3 or fewer 'heads' is also 0.073 and the evidence in favour of the alternative hypothesis is now 0.146. Thus there is no longer support in favour of bias, and we would not reject the null hypothesis at any of the fixed significance levels. (It can be noted in passing that we use quotation marks everywhere to distinguish 'tails' when they are outcomes of spinning coins from tails when used in connection with distributions and tests.)

We therefore note that a one-tail test applies when we have an alternative hypothesis whose values all lie *on one side* of the null hypothesis value ($H_0: \theta = 0.5$ versus $H_1: \theta > 0.5$), while a two-tail test applies when the values in the alternative hypothesis lie on *both* sides of the null hypothesis value ($H_0: \theta = 0.5$ versus $H_1: \theta \neq 0.5$). The same applies for continuous measurements. If the manufacturer of camshafts had set the production process to produce camshafts of average length of 600.00 mm, then some time after the process had started he might wish to check that nothing had altered in the settings and so would like to test $H_0: \mu = 600.00$ versus $H_1: \mu \neq 600.00$. If he collected exactly the same data as given earlier, and if the standard deviation of the population is again known to be 1.5, then he would again have a value $z = 0.947$. But now, the process could have changed in one of two ways—either producing camshaft lengths greater than 600.00 on average, or ones less than 600.00 on average. Values more extreme than 0.947 would be ones greater than 0.947 in the former case, or less than −0.947 in the latter case (because if the sample mean is less than 600.00, then z will be negative). So the significance level this time is $P(z \geq 0.947) + P(z \leq -0.947)$. The symmetry of the standard normal distribution shows that the latter probability is equal to the former, so the significance level becomes $0.172 + 0.172 = 0.344$. Thus there is no evidence at all that the production process has slipped in any way.

Choice of statistic for the test

In the two examples used to introduce the ideas of hypothesis testing above, we have simply presented the statistics on which the tests are based (usually called the *test statistics*) as either 'reasonable' or 'obvious' ones to take, but how should we tackle the choice of a test statistic in general? To make progress, we first need to list the essential characteristics of a test statistic. On examining the examples above, we see that there are two vital ingredients. First, the test statistic must behave differently when the null hypothesis is true from how it behaves when the alternative hypothesis is true, in order for it to be able to distinguish between the two cases. Thus, in the coin spinning example we would expect the number of 'heads' to be somewhere in the middle of the range when the coin is unbiased, but closer to the extremes of the range when it is biased. Likewise, we would expect z to be somewhere near zero if the population mean of camshaft lengths is equal to 600.00, but somewhere much further from zero if the population mean departs from 600.00 in either direction. So the first condition is satisfied in our examples. Second, since the logic of hypothesis testing requires us to find probabilities of getting extreme results when the null hypothesis is true, the sampling distribution of the test statistic must be known *exactly* when the null hypothesis values are substituted for any unknown parameters, so that required probabilities can be calculated (from statistical tables if necessary). In the coin tossing example the distribution of number of 'heads' is the binomial with n specified by the number of spins and $\theta = 0.5$ when the null hypothesis is true, while in the camshaft lengths example, z has a standard normal distribution when the null hypothesis is true, and it has been seen that numerical probabilities are obtained easily in both cases. So the second condition is also satisfied in these examples. However, when faced with an unfamiliar hypothesis testing problem, how can we home in on an appropriate test statistic?

This is another area in which Sir Ronald Fisher made an important contribution in the 1920s, when he defined what he termed a *sufficient statistic* for an unknown population parameter. In layman's terms, a sufficient statistic is a single statistic obtained from a sample of values that encapsulates all the information contained in the sample about the parameter. To understand this statement fully, we must be slightly more technical and we must explore what we mean by 'information about a parameter'. Suppose that we have a sample of measurements from a population, and that θ is a parameter of the model we have chosen for the population. The sample contains information about θ because the distribution of the data, and hence of any statistic obtained from the data, involves θ so can be used as a basis for estimating θ. Thus the sample of camshaft lengths contains information about the population mean μ, as we have already demonstrated by constructing confidence intervals for μ from the sample. Slightly more technically, we saw in the first chapter that we can calculate conditional probabilities of one event given the occurrence of another, and we can apply the same idea to sampling distributions. In particular, we can calculate the distribution of the sample data given that a particular sample statistic S (say the sample mean) has a particular value. Such a conditional distribution is the set of all possible samples that lead to the given value of S, so it describes the possible variation in the sample data consistent with S having the fixed value—namely, the variability that has not been accounted for by S. Such conditional distributions will also in general depend on the parameter θ, which means that there is *further* information in the data after that due to S has been 'used up'. However, if the conditional distribution of the data for fixed value of S does not involve θ, then there is no more available information about θ, and S *is* a sufficient statistic for θ.

So a sufficient statistic is one that, in some sense, extracts the maximum amount of information about θ from the data, and it makes sense to use such a statistic when constructing a test of hypothesis. There are various techniques for finding sufficient statistics corresponding to any given model, based on the likelihood of the data. Moreover, a very useful theoretical result is that the maximum likelihood estimate of a population parameter is always a function of a sufficient statistic for that parameter, so use of maximum likelihood will generally lead to useful test statistics. The problem is, of course, in determining the sampling distribution of the resulting statistic, and often recourse has to be made to large sample results and the central limit theorem for approximations. Fortunately, in most standard situations the appropriate test statistic turns out to be the same as the corresponding pivotal function for confidence interval construction, but with the null hypothesis value replacing the unknown parameter in this function. The appropriate sampling distribution for finding significance levels is the same as the distribution of the pivotal quantity, so this enables tests to be conducted very simply. This is the basis of t-tests for population means, chi-squared tests for population variances, and F-tests for ratios of variances. Thus if we use the sample standard deviation 1.34 instead of the assumed population value of 1.5 in the calculation of z, and if the sample size is small, then we would need to refer the calculated value of z to the t-distribution on the appropriate number of degrees of freedom in order to obtain the significance level. Full details of this test, and all other standard tests, may be found in most introductory textbooks.

Some consequences; the power of a test

Let us return to the camshaft length example where the manufacturer wished to test H_0: $\mu = 600.00$ versus H_1: $\mu > 600.00$, and consider the population standard deviation σ as known to be 1.5. Suppose we decide to test at the 5% level of significance, so that we will reject the null hypothesis in favour of the alternative if the calculated value of $z = (\bar{x} - 600.00)/(1.5/\sqrt{n})$ exceeds 1.645 (where $\bar{x}$ is the mean of a sample of n camshaft lengths, and 1.645 is the value that is exceeded 5% of the time when drawing observations from a standard normal distribution). What are the consequences of this procedure, if we were to take repeated samples of size n from the same camshaft length population?

The consequences can be divided into two parts—those when the null hypothesis is true and μ is indeed 600.00, and those when the null hypothesis is false (i.e. the alternative hypothesis is true) and μ is greater than 600.00. Let us take the former case first. If $\mu = 600.00$ then z *does* have a standard normal distribution, so if it is calculated for each of a large number of repeated samples from the same population the value 1.645 will be exceeded on roughly 5% of occasions. But we have decided to reject the null hypothesis when z exceeds 1.645, so it follows that all such rejections will be *wrong*. In other words, whenever we decide to adopt the reject/do not reject approach at the 5% significance level, we are admitting that there is a 5% chance we have rejected the null hypothesis *incorrectly*. The same reasoning will clearly hold whatever the level of significance we choose—10%, 5%, 1% or any other. The significance level is actually the probability of making the mistake of wrongly rejecting the null hypothesis. So, we might argue, it is better to test at the 1% level since that way we reduce the chance of this mistake from 5% to 1% (the critical value rises from 1.645 to 2.326, so fewer samples exceed this new value). Continuing the argument, would it not be better still to test at the 0.1% level? Or even virtually eliminate the chance of error by testing at the 0.0001% level?

The basic flaw with this reasoning is that when we make it harder to reject the null hypothesis incorrectly, by pushing the critical value for z further and further out into the tail(s) of the standard normal distribution, we automatically make it harder to reject the null hypothesis when we *should* be rejecting it, namely when the alternative hypothesis is true. So although we are reducing the chance of making one mistake (incorrectly rejecting H_0), we are increasing the chance of making another mistake (incorrectly *not* rejecting H_0). We need to strike a balance between the two mistakes (often called the type I and type II errors, respectively), and this balance is one of the reasons why the traditional significance levels of 5% and 1% have become popular.

To examine this situation in a little more detail, let us turn to the second case mentioned above and look at the consequences for the camshaft lengths when the alternative hypothesis is true. For definiteness, let us stick to samples of size $n = 100$ (so that $\sqrt{n} = 10$) and continue with the known population standard deviation case of $\sigma = 1.5$ (so that $\sigma/\sqrt{n} = 0.15$). Since the alternative hypothesis is true, let us assume that $\mu = 600 + a$, for some value a. This will therefore also be the mean of the sampling distribution of $\bar{x}$. The upshot of all this is that z will still have a normal distribution, but now with mean $a/0.15$ and standard deviation 1, instead of mean zero and standard deviation 1. This implies that values of z are shifted along, and will be greater on average by $a/0.15$ than when the null hypothesis is true. So when the alternative hypothesis is true, values of z (over repeated samples from the same population) will be pushed towards the upper tail of the standard normal distribution, and hence the null hypothesis is more likely to be rejected. In fact, if we are testing at the 5% significance level, the probability of rejection becomes the probability that the standard normal distribution exceeds the value $(1.645 - (a/0.15))$. Clearly, this probability increases as a increases; some representative values are 0.05 at $a = 0$ (obviously!), 0.095 at $a = 0.05$, 0.164 at $a = 0.1$, 0.228 at $a = 0.15$, 0.64 at $a = 0.3$, 0.955 at $a = 0.5$ and 1.000 to 3 decimal places once a exceeds 0.75.

The probabilities calculated above are the probabilities of rejecting the null hypothesis when it *should* be rejected, for different values of μ within the alternative hypothesis. For a given value of μ, this probability is known as the *power* of the test at that value of μ. We can plot the values of the power for the different values of μ, and the resulting curve is called the *power curve*, or alternatively the *power function*, for the test. The typical shape of any power function is a sigmoid (i.e. flat at the base, then curving upwards initially steeply but levelling off and eventually flat), starting at the significance level of the test for the null hypothesis value and reaching the asymptote (final flat bit) at probability value 1.0. The power curve for the camshaft lengths above is shown in Figure 4.10. In general, the speed at which the curve reaches the asymptote is an indicator of how 'good' the test is: a very steep curve shows that most values in the alternative hypothesis lead to a high probability that the null hypothesis is rejected, but a very gradual rise shows that for many values in the alternative hypothesis there is a high probability that the null hypothesis will not be rejected. If there are two (or more) competing tests for the same hypotheses, then they can be compared by comparing their power curves. If the curve of one test always lies above the curve for another test, then the first test is said to be *more powerful* than the second; if the curve of one test always lies above the curves of all other competing tests, then that test is said to be the *(uniformly) most powerful* test. This is clearly the 'best' test that can be found for the given hypotheses, because it gives the best chance of rejecting the null hypothesis in those situations in which it should be rejected.

Some connections and interpretations

The reader may well have been struck by the connections that have emerged between the tests discussed above and the confidence intervals derived from

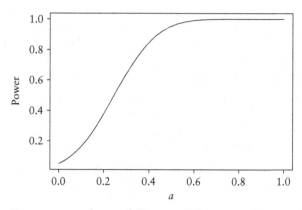

Figure 4.10.   Power curve of test of $H_0: \mu = 600.0$ versus $H_1: \mu > 600.0$ for the camshaft lengths.

the same samples in a previous section, in particular the use of the pivotal function from confidence intervals in constructing the test statistic. This is indeed no coincidence, so let us delve a little deeper into these connections.

Recollect first the calculation and interpretation of a 95% confidence interval for the mean μ of the population of camshaft lengths: over a large number of samples of size n from this population, the mean μ will lie between $a = \bar{x} - 1.96(\sigma/\sqrt{n})$ and $b = \bar{x} + 1.96(\sigma/\sqrt{n})$ on 95% of occasions, where $\bar{x}$ is the sample mean and the population standard deviation σ is assumed to be known.

Now consider setting up a test at the 5% significance level of $H_0: \mu = \mu_0$ versus $H_1: \mu \neq \mu_0$, where μ_0 is some specified value, using a sample of size n. Following through the standard procedure for a two-tail test on this population (normal model and σ assumed to be known), we have to reject the null hypothesis if the calculated value of $z = (\bar{x} - \mu_0)/(\sigma/\sqrt{n})$ is either greater than 1.96 or less than -1.96. Thus we will *not* reject the null hypothesis if z lies between -1.96 and 1.96. Carrying out a simple bit of algebraic manipulation on the right-hand side of the equation for z, this is exactly equivalent to saying that the null hypothesis is *not* rejected if μ_0 lies between $\bar{x} - 1.96(\sigma/\sqrt{n})$ and $\bar{x} + 1.96(\sigma/\sqrt{n})$. Thus we *do not* reject the null hypothesis at 5% significance if the hypothesised value lies *inside* the 95% confidence interval for μ, and we *do* reject it if the hypothesised value lies *outside* this interval.

The connection between the test and the confidence interval is entirely reasonable at an intuitive level. Over a large number of samples, the mean μ will fall outside the interval on 5% of occasions. So if our null hypothesis is true, then there is a 5% chance that it will fall outside our calculated interval and this is the same as the 5% chance that we reject the null hypothesis. This connection holds between any hypothesis concerning a parameter θ against a *two-sided* alternative at the $p\%$ significance level and the $(100 - p)\%$

confidence interval for θ, for any value of p between 0 and 100. However, it must be stressed that the alternative hypothesis has to be two-sided, so the corresponding test is two-tail, for the connection to hold exactly.

Nonparametric tests

In all the procedures above, it is necessary to assume some specific probability model for the population or populations about which the inference is required, and the outcomes of the inferential process (e.g. confidence intervals or significance levels) depend crucially on this assumption. The corollary to this is that if we have made the wrong assumption, and some other probability model correctly specifies the population, then our conclusions may be grossly in error. Once the frequentist approach to inference had gained some measure of acceptance, statisticians began to worry about the crucial nature of this assumption, and to develop techniques that did not require it. The goal was to provide methods of estimating population quantities or testing hypotheses about them that did not require explicit probability models for the populations. Since such probability models involve parameters in their specification, the methods described above became known as *parametric* methods, and so by contrast any methods that did not involve assumption of specific probability models became known as *nonparametric* methods. Of course, the focus of inferences to be made is still on population quantities (i.e. parameters), but no longer ones linked to specific models, so an alternative title sometimes given to these methods is *distribution-free*. However, irrespective of title, the frequentist framework of inference is fully retained in all these methods.

The past 50 years has seen the development of many such nonparametric methods, so the reader needs to refer to a specialist text for details. Our purpose here is just to give a flavour of the sort of approach undertaken, and to provide one or two general comments about the methodology. We shall focus exclusively on hypothesis testing, and more specifically on tests of hypotheses about population locations.

The general approach in the absence of particular probability models for the data is to replace the actual data values by some simplified quantities that have intuitive appeal with regard to the hypotheses of interest. The two most popular quantities of this type are *signs* and *ranks*, so we shall give some illustrative examples of each.

The simplest situation is a test of hypothesis that a population quantile has a specified value, because this defines the probability that a randomly chosen observation from the population will be less than (greater than) the specified value. The most common situation would probably concern a population median, so to consider a concrete example let us test the hypothesis that the median of the population of camshaft lengths is 600.9.

If the population median is indeed this value, then the probability of a randomly chosen camshaft length being less than 600.9 is 0.5, and likewise the probability it is greater than 600.9 is also 0.5. If we denote all sample camshaft lengths that are less than 600.9 by a minus, and all that are greater than 600.9 by a plus, and count the number of minuses we have, then we are in essentially the same situation as for the coin tossing example. We let θ be the probability of a randomly chosen camshaft length being a minus, and then test $H_0: \theta = 0.5$ versus $H_1: \theta \neq 0.5$ using the earlier method. This is equivalent to a test of the null hypothesis that the population median is equal to 600.9 against the alternative that it is not. The same idea extends to a test of a quartile or decile of the population, the only change being to the assignment of plus and minus and the null hypothesis value of θ.

This specific example raises an interesting point concerning the precision of the stated hypothesis. Suppose we test the hypothesis that the population median is 600.85 instead of testing that it is 600.9. Since the data are only recorded to the nearest 0.2 of a millimetre, exactly the same number of pluses and minuses are obtained as before so exactly the same test and conclusions follow from them. In fact, we will get exactly the same test and conclusions if the hypothesised median m lies anywhere in the range $600.8 < m < 601.0$, so any conclusions should relate to the median in this range. A similar consideration arises in any nonparametric test that involves a categorisation (of any form) of the data; the hypothesis and the conclusions should always take into account the precision of the original data.

If we want a nonparametric hypothesis test about a population mean, the situation is a little more involved. If the population is symmetric, of course, then the median and mean coincide. So one possible test is for a symmetric population plus a specified value of the mean. In this case, simply categorising values as plus or minus is not sufficient, because it conveys no information about their symmetry. To incorporate this information, we need to first calculate the deviations of each value from the hypothesised mean, rank all these deviations according to their magnitude only (i.e. ignoring any minus signs that exist), and then attach a plus or minus to the ranks depending on whether that observation is bigger than or less than the hypothesised mean. If the hypothesised mean *is* correct, then we would expect a good scatter of positive and negative ranks, so the sum of the positives should be close to the sum of the negatives. A big discrepancy suggests an excess of ranks carrying one of the signs. This would indicate rejection of the null hypothesis, and there are tabulated values to give guidance on how big this discrepancy needs to be in order to reject at a particular significance level.

A third situation along these lines is when we have separate samples from two populations, and we are interested in testing the hypothesis of equality of the median in the two populations. Here the null hypothesis is simply that the two populations have the same (unspecified) distributions, which of course implies equal medians. If this is indeed the case, and we rank the full set of observations from the combined samples, then we would again expect a good mixing of the ranks between the two sets. So the sum of the ranks for observations in one sample should be similar to the sum of the observations in the other sample. However, if the two populations have the same distributional forms (again unspecified) but their medians differ, then we would expect the ranks of one sample to be generally higher than the ranks of the other, and there would be a large discrepancy between the two sums of ranks. Again, there are tables available for a judgement to be made as to how large the discrepancy must be for the null hypothesis to be rejected at a given level of significance.

These three brief examples perhaps indicate the seductiveness of such nonparametric methods. They jettison restrictive assumptions about probability models, so they are applicable with confidence to a wide variety of data types, and they have an attractive intuitive basis allied to simple mechanical calculations. Little wonder that they enjoy a considerable following in many substantive areas. However, a brief cautionary note is perhaps in order. The downside to greater generality and simplicity is the danger of a loss of power in the tests. Whereas the user controls the type I error by fixing significance levels at desired values, so that the probability of wrongly rejecting the null hypothesis is known, the probability of correctly rejecting the null hypothesis can be very low. In general, there is a decrease in power whenever sample information is jettisoned, and in the various nonparametric tests this happens because the data values are replaced by less precise information. The extreme case occurs in sign tests, where continuous measurements are replaced by binary values (+ or −). Thus a sign test is likely to be less powerful than a corresponding test based on ranks, which retain information about the ordering of the original data values. Taking this argument to its logical extreme, the most powerful test will be one that retains all the original data values, but this requires a parametric test. So a parametric test should always be used in preference to a nonparametric one *if there is confidence that the assumed population model is appropriate.*

It is therefore appropriate to conclude this section on nonparametric methods, and the chapter on frequentist inference, by mention of some standard tests of goodness of fit of data to proposed probability models. If a set of data can be shown to be consonant with a particular population model, then the relevant parametric technique can be employed for maximum power. Moreover, the standard goodness of fit tests are essentially nonparametric in nature, so belong naturally in this section. All the tests measure some aspect of divergence between the proposed model and the corresponding sample quantity. The Kolmogorov test can be used to decide whether a sample comes from a specific probability distribution, and it does so by using as test statistic the maximum discrepancy between the population and sample cumulative distribution functions. There are tables against which to judge whether or not the proposed model should be deemed appropriate. On the face of it, this would be a good way of checking whether the normality assumption is adequate for a set of data, but unfortunately the Kolmogorov test requires the parameters μ, σ of the normal distribution to be known. A variation of this test, the Lilliefors test, is therefore often used. In this test the data are first standardised using the sample mean and standard deviation, the maximum discrepancy is then found between the standardised sample and standard normal cumulative distribution functions, and the resulting value is referred to a separate table to decide on model adequacy.

But perhaps the best-known test of adequacy of a model is the Pearson goodness of fit test, often called the chi-squared goodness of fit test. This test was designed principally for discrete measurements, and for each value r of this measurement it compares the observed frequency O_r with the frequency E_r that is expected under the given model. The discrepancy between the two sets of frequencies is measured by

$$X = \sum_r \frac{(O_r - E_r)^2}{E_r},$$

with large values of X indicating poor correspondence between the two sets and hence an inappropriate model. Tables of the chi-squared distribution need to be consulted in order to determine whether or not the postulated model is adequate. There are various technical issues involving degrees of freedom and pooling of categories that need to be addressed with this test, which we do not go into here, and it may also be noted that the test can be adapted for continuous measurements by using the observed and expected frequencies in a grouped frequency distribution. A comparison of the actual and theoretical distributions of defects on 400 pieces of cloth given in Table 3.3 (but without rounding the latter frequencies, and after resolving the discrepancies mentioned in Chapter 3) gave the value $X = 0.48$ on 4 degrees of freedom. This is an extremely small value, which indicates that the Poisson model is a very good one for this set of data, and this conclusion is confirmed following reference to the chi-squared tables. On the other hand, if the camshaft frequencies of Table 2.3 are tested for consonance with a normal distribution we obtain the very large value of 32.4 of X on 2 degrees of freedom, showing that a normal model is not appropriate for the data—which is not surprising given the histogram of Figure 2.1.

This last result should ring a warning bell regarding the uncritical application of inferential techniques that depend on the assumption of normality of the data. Of course, the central limit theorem can be invoked to ensure large-sample approximate normality of the sample mean whatever the parent distribution, so this theorem provides a justification of some inferences concerning the population mean (as in the case of the confidence intervals for, and tests about, the mean of the camshaft population above). Also, nonparametric tests are always available for population quantiles, including the median. However, if more powerful tests are required for the population mean, or if inferences are sought about other population characteristics, then it is essential that appropriate models are chosen for the relevant populations. It should be stressed that the formulation of such models is not generally a one-off theoretical exercise, but usually involves an iterative process of model proposal, testing of fit, and model refinement until a satisfactory outcome is achieved. It should also be self-evident, but is often forgotten, that this process should involve close cooperation and frequent discussion with the investigator who collected the data! The bottom line is that the quality of any inference is only as good as the model on which it is based.

5 Statistical Inference— Bayesian and Other Approaches

Introduction

Most of the basic structure of frequentist inference, as outlined in the previous chapter, was formulated in the period between the two world wars. Although it rapidly became established as the framework for general statistical analysis, and although theoreticians continued to produce important advances within this framework throughout the 1940s and 1950s, many statisticians were unhappy with it and many researchers either misunderstood or misinterpreted the techniques they were using. The root of the problem, as far as both groups of individuals were concerned, resides in the fact that results have to be interpreted in terms of long-run frequencies of occurrence of sample values. Statisticians were unhappy because they wondered why outcomes that had not occurred but were only potential ones in future hypothetical samples should have any bearing on the single actual outcome under consideration. Researchers misinterpreted frequency-based statements about observations as probability statements about parameters or hypotheses and hence misunderstood the conclusions drawn from the techniques.

For example, we have seen that a 95% confidence interval for a parameter θ is actually a recipe that will produce a (random) interval containing (the fixed) θ on 95% of occasions, if computed for each of a very large number of samples from the same population. This is not the same as saying that there is a probability of 0.95 that (a random) θ lies within the (fixed) interval, but the latter interpretation is the one that is often taken. Worse misinterpretation can arise with hypothesis tests. We have seen that the significance level simply tells us how frequently we will obtain a result at least as extreme (in the direction of the alternative hypothesis) as the one we have observed if the null hypothesis is true and we were to take very many samples from the same population. Yet the significance level is often interpreted as being 'the probability that the null hypothesis is true'. These two examples are common specific misunderstandings, but the more fundamental objection about frequentist inference in general can be paraphrased by the complaint 'We are only concerned with the sample in front of us; why should we be interested in what *might* happen if we were to take lots of samples from the same population?'

The upshot of this debate, and of the difficulties in correct practical interpretation, was that almost as soon as the frequentist approach became established, statisticians began to look for alternative approaches. The solution to both the problem areas outlined above seemed to lie in adopting a degree of belief interpretation of probability, so that any inference could be interpreted validly as a strength of belief about a particular statement, and this led naturally on to a Bayesian approach. So Bayes' theorem, which had existed quietly and uncontroversially as a useful result in probability theory for nearly two centuries, suddenly found itself pitched into the arena of statistical inference. To many it appeared to be the perfect answer to all the objections to frequentist methods, and they espoused it to the exclusion of everything else. Diehard frequentists, however, and anyone who did not subscribe to a degree of belief interpretation of probability, opposed this new approach equally fiercely at all opportunities. The early years of Bayesian methods were thus turbulent ones. The cause was not made any easier by technical mathematical problems that made many results difficult to calculate, which meant that approximations and unrealistic assumptions were necessary if useable quantities were to be derived—thus providing more ammunition for the frequentist. Fortunately, the rapid computational advances that occurred towards the end of the century have revolutionised the application of Bayesian methods, have removed the need for approximation, and so have brought this approach into (almost) universal acceptability.

The initial polarisation in the statistical community that the frequentist/ Bayesian controversy provoked led those individuals who held the middle ground to try and devise approaches that either drew upon or provided a compromise between the opposing camps. Some of these attempts were short lived, such as Fisher's idea of fiducial inference, which was a rather complex attempt to provide an interpretation of probability that blended frequentist and degree of belief ideas. Others, such as likelihood inference, have survived but within fairly restricted confines. Yet others, principally decision theory, were essentially different frameworks but ones which nevertheless drew upon Bayesian ideas for their execution. The purpose of this chapter is to give the reader a flavour of these various different approaches to inference. We will concentrate in the main on the Bayesian approach, as this is now the main competitor to the frequentist approach, but towards the end of the chapter we will briefly highlight some of the features of these other approaches. Much of the controversy surrounding the various approaches hinges on technicalities that cannot be dealt with satisfactorily in a book such as this, but nevertheless it is hoped that the reader will gain some overall appreciation of the differences between the various approaches.

Bayesian inference

The Bayesian approach

Since we are always concerned in inference with making statements about a population or populations from sample data, the probability model for the population retains a central place and (keeping to a parametric approach) any inferential question is translated into one about the parameter(s) of the model. In the frequentist approach we assume the parameter(s) to be fixed and unknown, and all probabilities to relate to the data in a relative frequency sense. This approach therefore focuses exclusively on the data, and makes no provision for any prior information or intuition that the investigator may have about the model parameters. However, in the Bayesian approach we allow the unknown parameter(s) also to have a probability distribution, but interpreted in the sense of degrees of belief, so that the researcher can express his or her knowledge about their possible values. Although the two probability distributions (for data and model parameters, respectively) differ in the way in which they are interpreted, they can be combined and manipulated using standard rules of probability. The data can thus be used to modify these degrees of belief, and all inferences about the parameters are then based on the modified probability distributions.

To give some flesh to these ideas, suppose that we have a batch of n electric light bulbs, each having a fixed but unknown wattage θ about which we wish to make an inference. To do this we determine the lifetimes $t_1, t_2, \ldots, t_n$ of the bulbs, and formulate a suitable probability model for the lifetime t of a bulb of wattage θ—say the exponential model with mean depending on θ. The frequentist approach would then proceed by either calculating a confidence interval for θ or conducting a hypothesis test for θ given the observed lifetimes. However, there is no room in this approach to include any prior information that we might have about θ, such as that it is highly likely to be one of the standard wattages 40, 60, 100, 150 or 200. So a confidence interval, for example, pays no regard to this information but is simply a continuous interval which may or may not include several of these values. In a Bayesian approach, on the other hand, the investigator can build such information into the process by specifying a probability distribution for the possible values of θ. Suppose, for example, that the investigator believes that the only possible values are the five standard ones, and that there is no reason to prefer any one of these values to any other. Then the probability distribution for θ would consist of the probability 0.2 for each of these five values of θ. The Bayesian approach then involves calculating how these probabilities are modified in the light of the observed data values $t_1, t_2, \ldots, t_n$, and making inferences in the light of these modified probabilities.

In order to develop the appropriate structure and methodology, we must first express these various steps in general terms and with suitable notation to avoid any ambiguity. The probability model for the population is a function that gives us information about the probabilities of different outcomes of our measurements. If we denote the measurement generally by x, then the probability model specifies either the probability of obtaining a given value of x when the measurements are discrete or the probability density function if the measurement is continuous. Let us denote either of these cases simply by the function $f(x)$ (which was the notation used in Chapter 3 for the probability density function). But the probability model includes one or more parameters in its specification, so we must also indicate this fact in our notation. For simplicity let us denote the parameters by the single symbol θ; if there is only one parameter then it is θ itself, while if there are more than one then θ will stand for the collection of parameters. Since the probability model is fully determined once we know the value of θ, it is appropriate to indicate the presence of θ in the model by the conditioning symbol, so that $f(x \mid \theta)$ will denote our population model symbolically.

Next we turn to the probabilities attached to the parameters. These are different in type from the probabilities given by the population model, since they represent degrees of belief about θ while the probabilities generated from the population model represent the relative frequencies of different measurement values over repeated sampling from the population. To distinguish these two essentially different types of probability, let us denote the probability function for θ values by $p(\theta)$. This will represent distinct probability values if θ is discrete (i.e. if only distinct values of θ are possible), or a continuous probability density function if θ is continuous (i.e. if θ can take any value in a continuum). Of course, we hit the same problems in the continuous case as we did with the more traditional probability density functions as discussed in Chapter 3. The probability that θ takes a specified value is actually the area under the curve traced out by the density function $p(\theta)$ over a very small interval surrounding θ. This means that values greater than 1 are entirely possible for $p(\theta)$, as indeed they were for any previous density function $f(x)$—when such a value is combined with an interval of very small width the area becomes less than or equal to 1, which represents a genuine probability value. For this reason it is rather loose terminology to talk of $p(\theta)$ as a 'probability', but we will do so because it helps to provide an intuitive explanation of some of the results that we will encounter. However, the true situation as briefly outlined above should always be kept in mind.

Since $p(\theta)$ represents an individual's degree of belief, it can change depending on circumstance. In particular, a degree of belief is almost certain to be influenced by any data that have been seen. So it is customary to write $p(\theta)$ as the degree of belief about θ *before* any data have been seen, and to write $p(\theta \mid x)$ as the degree of belief about θ *after* observing the data value x. By extension, if D denotes the complete set of data in a sample, then the degree of belief about θ after observing the complete sample is written

$p(\theta \mid D)$. To distinguish these cases, $p(\theta)$ is termed the *prior probability* (or *prior distribution*) of θ, and $p(\theta \mid D)$ is termed the *posterior probability* (or *posterior distribution*) of θ.

The central tenet of Bayesian inference is then simple to state: any inference about θ is based exclusively on the posterior distribution $p(\theta \mid D)$, which expresses all the available information about θ. The first step is therefore to determine how this posterior distribution can be calculated. Recollect Bayes' theorem from Chapter 1: if A and B are two events, then

$$P(B \mid A) = \frac{P(A \mid B)P(B)}{P(A)}.$$

Let us therefore replace events A and B by the data D and the parameter(s) θ, respectively, and in place of single probabilities P let us have probability distributions p or f. Then direct substitution of these quantities into the formula above yields

$$p(\theta \mid D) = \frac{f(D \mid \theta)p(\theta)}{f(D)}.$$

Assume for the moment that we have specified the prior distribution $p(\theta)$ in a particular case, as in the light bulb example above, and that we have formulated a suitable population model $f(x \mid \theta)$ for the data. Since the observations in a random sample are independent of each other, the probability of the data $f(D \mid \theta)$ is obtained by multiplying together the values of the probability model at each measurement value in the sample. When x is a discrete quantity these values are the probabilies of the sample measurements, while if x is continuous then they are the probability density values at each sample measurement. We gave an intuitive argument in Chapter 3 as to why $f(D \mid \theta)$ formed in this way can be interpreted as the probability of the data, and termed it the *likelihood* of the data when it is considered as a function of θ. It has already been used to find the maximum likelihood estimate of a parameter θ, and indeed its various mathematical properties make it a familiar concept in frequentist inference. The above result means that it also plays a central role in Bayesian inference.

The denominator $f(D)$ in the above formula generally presents more problems, but we can avoid its explicit calculation. Since it does not involve θ, it remains constant for different values of θ so that as far as the posterior distribution (as a function of θ) is concerned we can write

$$p(\theta \mid D) = kf(D \mid \theta)p(\theta),$$

for some constant value k. To find what this value is in any specific case, we use the fact that $p(\theta \mid D)$ is a probability distribution. This means that if θ takes only discrete values, the sum of $kf(D \mid \theta)p(\theta)$ over these values must equal 1 and this determines the value of k. The corresponding condition when θ is a continuous parameter is that the total area under the curve

$kf(D \mid \theta)p(\theta)$ must equal 1. In principle this should also yield the value of k, but to obtain the total area we need to use integral calculus and the calculation is not always possible. So there is a technical difficulty here, which is one of the causes of previous problems and controversies with Bayesian inference. We will return to this point below.

At this juncture, it is worth giving a few simple examples to show these calculations in action. First suppose that there are two identical-looking coins, but one is unbiased so that a 'head' is as likely as a 'tail' when the coin is spun while the other is biased with a 'head' being three times more likely than a tail. In terms of probabilities, this means that $P(\text{'head'}) = 0.5$ for the unbiased coin, but $P(\text{'head'}) = 0.75$ for the biased one. To express this situation in the above general framework, denote the probability of 'head' by θ so that θ is a discrete parameter with possible value 0.5 and 0.75.

Suppose that I choose one of the coins purely at random. Then I have no prior preference between the two coins, so my two prior probabilities regarding the possible value of θ for the coin I have chosen, $p(\theta = 0.5)$ and $p(\theta = 0.75)$, are both equal and hence both are 0.5. If I now spin the coin and it comes down 'head', how should I re-express my beliefs about which coin I have spun? Well, here the data consist of the single outcome 'head' of the spin, and the probability of a head with either coin is θ. So $f(D \mid \theta) = \theta$, which is either 0.5 or 0.75 depending on the coin. Thus the two posterior probabilities are given by:

$$p(\theta = 0.5 \mid D) = k \times f(D \mid \theta = 0.5)\, p(\theta = 0.5) = k \times 0.5 \times 0.5 = 0.25k,$$

$$p(\theta = 0.75 \mid D) = k \times f(D \mid \theta = 0.75) \times p(\theta = 0.75) = k \times 0.75 \times 0.5 = 0.375k.$$

But since these two posterior probabilities must add to $1, 0.25k + 0.375k = 1$. Thus $0.625k = 1$, from which we find $k = 1.6$. Thus, finally

$$p(\theta = 0.5 \mid D) = 0.25 \times 1.6 = 0.4,$$

$$p(\theta = 0.75 \mid D) = 0.375 \times 1.6 = 0.6.$$

So, having at first had an equal degree of belief in each coin, the outcome of a 'head' has swung me towards greater belief (by 50%) in the biased coin over the unbiased one—an intuitively reasonable result.

To demonstrate the extra complication introduced by a continuous parameter, let us consider a similar situation in which I take a coin, spin it once, and observe a 'head'. But now suppose I know nothing about the potential bias of the coin and simply believe that θ, the probability of obtaining 'head', could be anywhere between 0 and 1. Thus my prior distribution $p(\theta)$ is now a uniform distribution with probability density $p(\theta) = 1$ over the range (0, 1). Once again $f(D \mid \theta) = \theta$, so now we have

$$p(\theta \mid D) = k \times \theta \times 1 = k\theta.$$

But the total area under this curve from 0 to 1 must equal 1. Either by using calculus, or from the geometry of the (triangular) shape of the density, we find this area to be $\frac{1}{2}k$, so that $k = 2$ and $p(\theta \mid D) = 2\theta$. This means that our posterior belief having seen the 'head' is weighted increasingly heavily towards higher values of θ, again an intuitively reasonable result but one that is harder both to derive and to interpret than in the simple discrete case earlier. It might also be noted in passing here that the posterior 'probability' $p(\theta \mid D) = 2\theta$ will be greater than 1 whenever θ is greater than 0.5. This illustrates the point made earlier, namely that the true probability of θ having a specified value is the area under 2θ of a very small interval surrounding the specified value, and that we only call 2θ the probability for ease of explanation.

Prior and posterior distributions: some mechanics

So we have seen how the posterior distribution can be derived, at least in principle, and once we have this distribution then we have all we need in order to make inferences about θ. But how do we set about formulating a prior distribution in order to start the process off, and can we somehow circumvent the technical problems associated with the computation of the constant k? These two questions are linked to some extent, so we consider them together. In all this discussion, we can omit any consideration of the discrete parameter case, as there are usually no problems here. The set of possible values of the parameter is restricted, so it is rarely difficult to come up with a set of probabilities that reflect our prior belief about these values. Having formulated the prior distribution, then the calculation of k is again based on only a limited set of values in the posterior distribution, so can be completed using basic algebra and arithmetic without difficulty. The problem areas arise when the parameter is continuous, so that is what we focus on here.

In general, we can break down the decision process associated with choice of prior distribution into two possible cases: (i) when we have no prior knowledge or information about the parameter and (ii) when we do have some prior knowledge. If we do not have any prior knowledge, then most individuals would argue that one should choose a 'neutral' prior, namely one that does not favour any single value or range of values of the parameter over any other value or range. This would generally suggest a uniform distribution over the range of possible parameter values, in other words a 'flat' density function having constant value c across the whole range of parameter values. This density function has the property that the probability the parameter is in an interval of specified width is the same, wherever that interval is located in the range of parameter values (which was the prior for the example above, and since θ had range 0–1 the appropriate value of c was 1). Such a prior distribution is commonly adopted for a location parameter (e.g. mean) in the absence of prior information. Unfortunately, an immediate problem appears if the location parameter can take any value from minus to plus infinity (which is the case for many location parameters, including the ones for a normal distribution), because with an infinite width we cannot find the correct height c to make the area under the density function equal to 1. Fortunately, this problem can be circumvented, because with this prior the constant c becomes subsumed into the previous constant k and the calculation proceeds as before. So such a prior is called an *improper prior*, because it does not satisfy the fundamental property of a density function but nevertheless is operationally satisfactory. Even if the prior distribution is improper, the posterior distribution can nevertheless be a proper probability distribution (since all constants introduced at any stage of the calculation are progressively combined and dealt with in one go at the end of the calculation). Such a uniform prior is adopted also for the logarithm of the standard deviation, and for many other parameters in the absence of prior knowledge.

However, there is another, slightly subtler, problem associated with the uniform prior distribution, and that is that it does not translate mathematically into a uniform distribution for an arbitrary function of the parameter. For example, if we choose a uniform prior for the standard deviation of the population, this does not translate mathematically into a uniform prior for the variance (i.e. squared standard deviation) of the population. Yet, if we have no knowledge about the standard deviation, then we might with justification ask where some knowledge about the variance has come from! Attempts have been made to overcome this problem, and one common approach to do so is to use the so-called 'invariant prior' developed by Harold Jeffreys, which is designed to be unaffected by such transformations; details of its use are given in texts dealing with Bayesian inference. Between them, the uniform and the invariant prior will cover most practical cases of prior ignorance about the parameter.

In the second case, where there is some prior knowledge, the usual advice is to use one of the standard distributions (such as the ones detailed in Chapter 3) in such a way as to adequately capture this prior knowledge. Each of these standard distributions has its own parameters, so values must be chosen for them in such a way that the resulting density curve represents the prior knowledge. Such choice requires practice and experience. However, some guidelines exist, and are particularly angled at making subsequent calculations simpler. We note that the prior distribution $p(\theta)$ is multiplied by the likelihood $f(D \mid \theta)$ as the first step in the derivation of the posterior distribution $p(\theta \mid D)$, and then the constant k is evaluated. If the choice of prior clashes with the likelihood in any way, then problems often arise in calculation of k. In the extreme case, this calculation becomes impossible. So to smooth the process, one looks for the most compatible functions for prior and likelihood. Since the population model determines the likelihood, it is natural to seek a function for the prior that is most compatible with the function for the population model. It turns out that for many standard models there exists what is known as a *conjugate family* of distributions from which the prior can be selected. Not only is such a prior entirely compatible with the likelihood, but the posterior distribution is then also guaranteed to belong to the same family of distributions so no awkward problems arise. Some examples of conjugate families are: the normal distribution for a normal population model, the beta distribution for a binomial population model, and the gamma distribution for either a Poisson or a gamma population model.

Standard inferences

The three standard inferential problems tackled in the frequentist approach are point estimation of the unknown parameter θ, interval estimation of θ, and hypothesis tests about θ, so let us now briefly consider how they might be tackled in the Bayesian approach.

The concept of point estimation is bypassed to some extent in Bayesian inference, because a single 'guess' at the value of the unknown parameter is rendered superfluous once one has the full posterior distribution $p(\theta \mid D)$ for that parameter. However, if a single value is needed for some reason then any of the usual single summary measures such as mean, median or mode of $p(\theta \mid D)$ can be employed. A rational case can be made in given circumstances for any of the three measures (and see also the discussion on Decision Theory at the end of this chapter), but perhaps the most intuitively appealing one in general is the mode as this value is the 'most likely' one with reference to our degree of belief in the parameter after seeing the data. It can also be noted that if we have little or no prior knowledge about θ and we use a 'flat' prior distribution for it, then we see from the earlier definition of $p(\theta \mid D)$ that

$$p(\theta \mid D) = kf(D \mid \theta)p(\theta) = Cf(D \mid \theta)$$

for some suitable constant C. So the posterior distribution is proportional to the likelihood of the data, and so these two quantities will have the same maximum over θ. In other words, the mode of the posterior distribution will coincide with the maximum likelihood estimator of θ in this situation. Of course, the two estimators will no longer coincide when either a conjugate or other informative prior is taken for θ in place of the flat one. The estimator provided by the mode of the posterior distribution is often referred to as the *maximum a posteriori (MAP)* estimator.

Much more useful in terms of parameter estimation is an interval estimate, and the Bayesian approach leads to an interval that is both intuitively reasonable and interpretable directly in terms of degree of belief. If we want an interval within which we believe the parameter to lie with probability α, then all we need to do is to find an interval (a, b) of values of θ such that the area under the posterior density curve between $\theta = a$ and $\theta = b$ is equal to α. Such an interval is called a $100\alpha\%$ *credible interval*. However, there are many intervals that satisfy this requirement for any given posterior density and value of α, so a further condition is needed in order to make the computation fully specified and the resulting interval unique. One possibility is to require the interval to be as short as possible for given α, and this is achieved by restricting the interval to contain only those values of θ for which $p(\theta \mid D)$ exceeds some constant c. Such an interval is called a *highest posterior density* interval. The problem with this type of interval is that it may actually consist of two disjoint parts if the posterior density has more than one mode (e.g. $1.7 < \theta < 2.9$ and $4.2 < \theta < 6.7$). This is not ideal, and moreover the computation can be awkward in some cases. The other way of fixing the interval uniquely is to apportion the remaining probability $(1 - \alpha)$ equally between the two tails of the posterior density, thereby achieving a *central* interval (also known as an *equal tail* interval). So, for example, a central 95% credible interval would have 2.5% of the density outside each end point and 95% of it inside. Such intervals are much easier to compute, accord with the traditional construction of the frequentist confidence intervals, and will coincide with the highest posterior density interval if $p(\theta \mid D)$ is both unimodal and symmetric. However, the interpretation of credible intervals is much more direct and intuitive than the interpretation of confidence intervals. They can now be correctly viewed as intervals within which the parameter lies with the given level of probability (i.e. degree of belief), and there is no repeated sampling mechanism necessary for this interpretation.

Hypothesis testing can be tackled equally directly in the Bayesian framework. Suppose that we wish to test the null hypothesis that θ takes one of the values in a specified set (or interval) of values represented by Ω_0 against the alternative that it takes one of the values in the set (or interval) of values represented by Ω_1 (i.e. $H_0: \theta \in \Omega_0$ versus $H_1: \theta \in \Omega_1$). Then the Bayesian approach is to calculate the two posterior probabilities $\beta_0 = p(\theta \in \Omega_0 \mid D)$ and $\beta_1 = p(\theta \in \Omega_1 \mid D)$ by finding the areas under the appropriate portions of the posterior density $p(\theta \mid D)$, and then to decide in favour of the hypothesis that has the greater β value. Note that if we write $\alpha_0 = p(\theta \in \Omega_0)$ and $\alpha_1 = p(\theta \in \Omega_1)$, then α_0, α_1 are our prior degrees of belief about the two hypotheses (before we have seen the data) while β_0, β_1 are our posterior degrees of belief about them (after we have seen the data). So the *prior odds* in favour of H_0 are α_0/α_1, while the *posterior odds* in favour of H_0 are β_0/β_1. Thus the ratio of these two odds, tells us what effect the

$$B = \frac{\beta_0/\beta_1}{\alpha_0/\alpha_1} = \frac{\beta_0 \alpha_1}{\beta_1 \alpha_0}$$

data D have had in modifying our prior beliefs about the two hypotheses. This ratio is known as the *Bayes factor* in favour of H_0, and is a measure of the odds in favour of H_0 against H_1 given by the data. Once again, though, all interpretations of these hypothesis tests can be related directly to probabilities as degrees of belief rather than through the cumbersome repeated sampling mechanism.

Predictive distributions

Arguably one of the most powerful uses of Bayesian methods is in the prediction of future events or measurements given some existing data. To show the basic steps of the methodology, let us revisit the very simple example we considered earlier in this chapter. There were two identical-looking coins, one unbiased with $P('head') = 0.5$ and the other biased with $P('head') = 0.75$. The probability of 'head' was denoted by θ, and I chose one of the coins purely at random so that the two prior probabilities regarding the possible value of the coin were $p(\theta = 0.5) = p(\theta = 0.75) = 0.5$. Having spun the coin once and noted that it came down 'Head', we showed that the two posterior probabilities were $p(\theta = 0.5 \mid D) = 0.4$ and $p(\theta = 0.75 \mid D) = 0.6$ Suppose now I want to predict the outcome of the next spin of the same coin. To do this, I first need to estimate the probabilities of the two possible outcomes, to see which outcome is more likely. Now the probability of a 'head' is θ, so if my coin is the unbiased one then I have a probability 0.5 that the next spin will land 'head' but if my coin is the biased one then I have a probability 0.75 that the next spin will land 'head'. So if I denote by B the probability of a 'head' on the next spin, then B is a discrete 'measurement' having the two possible values 0.5 and 0.75, and the probabilities of these two values are given by the two posterior probabilities above (since I now have the first spin of the coin as evidence, and these posterior probabilities summarise my degree of belief about θ). So I can summarise the distribution of B as in Table 5.1.

I can now conduct any of the usual operations on this distribution, and in particular I can find the expected value of B as $0.5 \times 0.4 + 0.75 \times 0.6 = 0.65$. This is therefore the expected (i.e. predicted) value of the probability that the next spin will land 'head'. We could now repeat the process for the probability that the next spin lands 'tail', simply by changing the values of B in Table 5.1 to 0.5 and 0.25 (probabilities of 'tail') but leaving $P(B)$ unchanged (since our posterior probabilities of θ values are unaffected). However, since there are only two possible outcomes, it is evident that the predicted probability of 'tail' must be $1 - 0.65 = 0.35$. Thus the *predictive distribution* of the outcome of the second spin, given the outcome of the first spin, is $P('head' \mid D) = 0.65$ and $P('tail' \mid D) = 0.35$. The odds on a 'head' are nearly 2 to 1 so we would predict 'head', but with the knowledge that there is a moderate (35%) chance of being wrong.

Table 5.1. Probability distribution of B (probability of 'head' on next spin)

$B \, [= \theta]$	0.5	0.75
$P(B) \, [= p(\theta \mid D)]$	0.4	0.6

Note that if we adopted a strictly frequentist approach, and eschewed any consideration of degrees of belief about parameter values, then we would be in a much weaker position even though the Bayesian approach started from a lack of prior knowledge. In the frequentist approach we need to estimate θ from the sample data alone. Since the data comprise just one spin of the coin, resulting in 'head', then our 'best' estimate is $\theta = 1/1 = 1$ and we would predict the probability of a 'head' on the next spin as 1—clearly a very poor prediction. The Bayesian approach has scored over the frequentist one by being able to average the possible values of the probability of a 'head' over the posterior distribution of the parameter, whereas the frequentist approach has had to take just one single estimate (and in this example that estimate is based on very skimpy data).

This general procedure can be followed for any predictive problem, no matter how complicated the quantity to be predicted (B in the above example) and regardless of whether the parameter is discrete or continuous. If the parameter θ is discrete, then the predictive distribution is obtained by summing the potential values of B over the posterior distribution of θ, while if the parameter θ is continuous then it is obtained by using integral calculus to find the area under the curve $Bp(\theta \mid D)$. Future predictions are then made on the basis of the probabilities in the predictive distribution. This procedure features centrally in a Bayesian approach to such predictive techniques as regression and classification considered later in the book.

As a tailpiece, we might note that modern computational approaches have revolutionised this area of predictive distributions, and have rendered many previously difficult problems perfectly tractable. The problem in the past was that, while the solution to most problems could always be formulated in terms of the predictive distribution, the necessary integration to compute this distribution was either very difficult or, often, impossible to carry out analytically. Frequently, the posterior distribution $p(\theta \mid D)$ could not be derived explicitly because of problems in computing the constant of proportionality. Moreover, when it could be derived the numerical integration ('quadrature') was often hampered because the functions in practical problems were high dimensional and hence very difficult to evaluate. However, the necessary integrals are simply expectations of B with respect to $p(\theta \mid D)$, so can be very closely approximated by averages of values of B in large samples from $p(\theta \mid D)$. The computational breakthrough came in a series of developments in the 1990s that showed how independent observations could be generated from any user-specified posterior distribution $p(\theta \mid D)$. This enables very large samples of B values to be generated from appropriate posterior distributions in virtually any practical application, and the predictive distribution can therefore be calculated as the mean of such a sample. The generic title for these computational methods is *Markov Chain Monte Carlo* methodology, and their wide availability in computer software packages has led to an explosion in the application of Bayesian methods to practical problems.

Fiducial inference and likelihood methods

When the Bayesian approach was first put forward as a viable alternative to frequentist inference, many statisticians who did not hold with the tenets of the latter could not reconcile themselves to the former either. They put forward two main objections, both centring on the use of prior information about the unknown parameter. The first was concerned with the inherent subjectivity of the concept, the objection being that different individuals would come to different conclusions from the same data and concomitant information if they formulated sufficiently different prior distributions. The introduction of a 'personal view' seemed contrary to scientific reasoning (to which the Bayesian counter was that there is often much experience and knowledge of a problem that should be included in the analysis as well as the data). The second objection concerned the forced imposition of a prior distribution when there was no prior knowledge about the parameters. Moreover, different theories about how such a prior distribution should be formulated could again lead to different conclusions from a single data set. So a number of attempts were made to come up with alternative frameworks. In this section, we briefly review two attempts at modifying Bayesian ideas to remove these objections, while in the next section we highlight the main points of an alternative approach that uses Bayesian ideas.

The concept of fiducial probability, first propounded by Fisher in 1930 as a measure of credence in a parameter value, has been one of the great enigmas of the twentieth century. Fisher's writing, while never transparent, excelled itself in its opacity on this topic. Moreover, matters were not helped by the fact that his viewpoint shifted over the 25 years following its first publication, and he backed it up with various arguments that later authors acknowledged as being erroneous. The consequence was that this was a topic very hotly debated during the years from 1930 to the early 1960s, but one which then virtually disappeared from public consciousness after Fisher's death in 1962. However, it is again being raised from time to time in connection with statistical method, so a brief mention here seems to be appropriate.

Fisher's purpose was to provide a mechanism whereby probability statements could be made about an unknown parameter, but without the contentious Bayesian requirement of a prior distribution for it. In his search for a solution, he unsurprisingly appealed to his previous discovery, the sufficient statistic, and linked it with the pivotal function used in the frequentist approach to confidence interval construction. So to describe his idea, let us go back to the simplest case we considered in Chapter 4—the confidence interval for a population mean μ, as illustrated by the camshaft lengths. We showed there that if we assume that a random sample of n values is drawn from a normal population having mean μ and *known* standard deviation σ, and that if $\bar{x}$ is the mean of the sample values, then the quantity $z = (\bar{x} - \mu)/(\sigma/\sqrt{n})$ is a pivotal function having a standard normal distribution. Even if the population is not normal but the sample size n is large, then we can use the central limit theorem to justify approximate normality of z (as we were able to do for the sample of $n = 100$ camshaft lengths).

Thus we can find values from standard normal tables that give the probability associated with statements about z, for example, we can say that the probability that z lies between -1.96 and $+1.96$ is 0.95. No one would argue with that statement, and the interpretation is that over many samples from a population having (fixed) mean μ, the calculated value of z will fall in this interval 95% of the time. Thus the statement is implicitly one about a *fixed* μ and a *variable* $\bar{x}$ (from sample to sample). Fisher, however, argued that this statement is equally true if $\bar{x}$ is assumed to be fixed while μ is assumed to vary and to have its own distribution—the *fiducial distribution*. In this example, therefore, the fiducial distribution of μ can be taken as normal with mean $\bar{x}$ and standard deviation $\sigma/\sqrt{n}$. Hence any confidence interval constructed from this pivotal quantity can be interpreted as a probability statement about μ.

This argument was propounded in somewhat elliptical fashion initially, and Fisher gradually came to assert that there was a perfectly legitimate probability distribution associated with unknown parameters, but this became a major source of contention and argument. Problems were compounded when it became evident that many situations existed in which pivotal functions of the required form either did not exist, or could not be inverted into a simple interval for the unknown parameter. Moreover, severe difficulties were encountered when there was more than one unknown parameter for inferences, and the combination of all these problems led to loss of interest in the method. However, before dismissing it completely, it is worth noting that there is a valid frequentist interpretation of probability statements about μ as made above. If we imagine all the possible normal populations with different means that produce a sample having the fixed value $\bar{x}$ for the sample mean, then the statement

$$P\left\{\bar{x} - 1.96\frac{\sigma}{\sqrt{n}} < \mu < \bar{x} + 1.96\frac{\sigma}{\sqrt{n}}\right\} = 0.95$$

carries a repeated sampling interpretation over these populations and hence is a justifiable probability statement about μ. So fiducial inference may yet creep back into the statistical mainstream.

An alternative way of circumventing the Bayesian controversy is to ignore the degree of belief approach entirely, and to base all inferences simply around the likelihood of the data $f(D \mid \theta)$. This is just the probability of the data, but viewed as a function of the unknown θ, so can be used as a measure of 'support' for different values of θ. We have already seen that the maximum likelihood estimator is the value of θ that maximises $f(D \mid \theta)$, so in this interpretation it is the value of θ that has the greatest support—say f_{max}. In order to standardise the support measure, many users prefer to work with the ratio $f(D \mid \theta) / f_{max}$, that is, the *relative likelihood* which lies between 0 and 1, and a system of inference has been built up around this ratio. For example, a *likelihood interval* for θ can be constructed from the set of values of θ whose ratios all exceed some critical value, and two hypotheses about θ can be assessed by comparing their ratios. Moreover, since the likelihood is itself a probability then concepts such as conditional and marginal likelihood follow directly, while extensions to partial likelihood allows more complicated multi-parameter problems to be tackled. This framework for inference has gained ground since the 1960s, and is implicitly present in various explanatory models in current use.

Decision theory

The rapid development in the 1940s of decision theory, namely the quantitative study of decision-taking in the face of uncertainty, was stimulated (as were many other branches of statistics and operational research) by the exigencies of the Second World War. It was initially formulated as a means of devising optimal strategies for deciding between specified courses of action, but was subsequently adapted to statistical inference in general. We therefore first give a brief indication of its basic features, and then consider how these fit into the more general situation.

Any decision theory problem will involve some or all of the following ingredients. (i) We need to decide between a number of 'actions', one of which we have to take at some point in the future. For example, we may want to decide whether or not to conduct some business tomorrow and our possible actions may be to go out with an umbrella, to go out without an umbrella or to stay at home. (ii) There are various 'states of nature' possible when we take our action, but at this stage we do not know what they will be. In the example, the simplest scenario envisages just two states of nature, viz. rain or no rain. (iii) We must be able to quantify the relative desirability or otherwise of each action for each state of nature. This can be done either via a 'loss function' that specifies the losses resulting from each action for each state, or via a 'utility' function that specifies the corresponding benefits. Specifying losses or utilities is often the most difficult and contentious aspect in a practical application. If all potential outcomes are strictly financial, then they can easily be equated to cost, but in most cases we also have to consider subjective factors such as quality of life, risk to health, effect on well-being. The theory usually focuses on loss rather than utility, so we will follow this custom. In our example above, we may judge there to be no loss if we go out without an umbrella when there is no rain, a minor loss (say 1 unit for inconvenience) if we go out with an umbrella when there is no rain, a major loss (say 6 units for cost of taxi) if we go out without an umbrella if there is rain, a medium loss (say 3 units for missing out on a good deal) if we stay at home irrespective of weather, and so on. (iv) Finally, we may also have some prior knowledge of the probabilities of the states of nature (e.g. today's weather), and some data to add to this knowledge (e.g. the weather forecast for tomorrow). Putting the two together using Bayesian precepts gives us posterior probabilities for the states of nature.

Formally, let us write a to denote the action (so in the example above this is a discrete quantity with values a_1: go out with umbrella, a_2: go out without umbrella, a_3: stay in), θ to denote the unknown state of nature (so in the example above we have θ_1: rain and θ_2: no rain), x to denote any relevant measurement and D to denote the data in a sample (in the example above this might be the weather forecast) and $p(\theta)$, $p(\theta \mid D)$ to denote as usual the prior and posterior probabilities of the states of nature (before and after observing the sample). Typically, there will also be available (either empirically or from a population model) the distribution $f(x \mid \theta)$ of probabilities of values of x for each θ. A *decision procedure* $\delta(x)$ specifies the action to be taken when we observe data value x (e.g. $\delta(x) = a_1$ if weather forecast (x) is 'rain tomorrow'), and the loss function $L(\theta, \delta(x))$ measures the loss from action $\delta(x)$ when the state of nature is θ.

A full specification for the solution of the decision problem will therefore require a table of values of loss for each combination of values of a and θ, the set of $\delta(x)$ procedures, and the probability distributions $p(\theta)$, $p(\theta \mid D)$ and $f(x \mid \theta)$. To find the solution, we first compute the mean (i.e. expected value) of the loss function with respect to the data distribution $f(x \mid \theta)$—this is known as the *risk function* and it will depend on θ and $\delta(x)$ (i.e. have different values for the different states of nature and the different decision procedures). The solution to the decision problem is based on the risk function; various routes are possible, but two have proved most durable in practice. The *minimax solution* is to choose that decision that will minimise the largest risk over all values of θ. This is usually considered to be the pessimistic route, as we are trying to minimise the 'worst that can happen to us'. A more positive approach is to first average the risk over the prior distribution $p(\theta)$, thereby converting it into the *Bayes risk*. The *Bayes solution* then chooses the decision that minimises this Bayes risk, which means that we are trying to minimise our 'anticipated' loss rather than our 'worst possible' loss.

To impose this structure on standard statistical inference, we simply associate θ with the unknown population model parameter rather than with the state of nature, and the decision procedure $\delta(x)$ with the inferential question we are trying to answer. So if we are estimating θ, then $\delta(x)$ is our estimator for given value of x, while if we are testing hypotheses then $\delta(x)$ is the decision to either reject or not reject the null hypothesis.

The estimation problem is perhaps rather more contrived as a decision problem, but nevertheless perfectly so adaptable. The problem here is that the unknown parameter is usually continuous rather than discrete, and there is some arbitrariness in the loss function that should be adopted. The objective is to find estimators that home in on the true value as far as possible, so the loss function should be designed to penalise estimators whenever they deviate from the true value. There are three such loss functions commonly considered: (i) *zero-one loss*, which allocates zero loss if the estimator is within a small distance a of the true value, and a constant loss of one unit otherwise; (ii) *absolute error loss*, in which the loss equals the absolute difference between the estimator and the true value; and (iii) *quadratic loss*, in which the loss equals the squared difference between the estimator and the true value. Following through the decision theory framework outlined above, it can be shown mathematically that adopting zero-one loss leads to the mode of the posterior distribution $p(\theta \mid D)$ as the optimal estimator; adopting absolute error loss leads to the median of $p(\theta \mid D)$ as the optimal estimator and adopting quadratic loss leads to the mean of $p(\theta \mid D)$ as the optimal estimator. Thus the decision theoretic framework not only potentially widens the choice of estimators (as there are potentially many other loss functions the user might entertain), but also gives a rationale for choices between Bayesian estimators that were previously made on arbitrary or intuitive grounds. For example, we see from above that the choice between mode, median and mean of posterior distribution reflects the progressive imposition of increasing penalty, the more that the estimator deviates from the true value.

The application of decision theory to hypothesis testing, while ostensibly much more natural than its application to estimation, leads to a much more technical discussion and so is not pursued here. It might be noted in passing, however, that decision theory plays a central role in *sequential* tests of hypotheses where, instead of taking a sample of fixed size n in order to test a hypothesis, the data are collected either one observation or a small number of observations at a time and the hypothesis is tested at each stage of the collection. The purpose of such a procedure is to ensure that the rejection or otherwise of the hypothesis is done as quickly and efficiently as possible, which is important in situations where the sampling is very costly, or where the decision carries heavy ethical considerations such as in clinical trials. As might be envisaged, decision theory is vital in setting up the mechanism whereby the stopping point of the process and the subsequent action can be decided optimally. However, any details of this process again take us into very technical material so are not pursued here either.

6 Linear Models and Least Squares

Introduction

The previous chapters have been mainly concerned with statistical *principles*, the focus being essentially on explaining the underlying logic of statistical reasoning while keeping mention of specific statistical *techniques* as somewhat incidental to this focus. For the second half of the book we turn more directly to statistical techniques, but the emphasis will still be on rationale and objectives rather than on technical details. There are now very many statistical tools available to the investigator, so that the inexperienced researcher may feel overwhelmed when trying to select an appropriate method for analysing his or her data. Our objective for this second half of the book is therefore to survey the most common and useful members of the statistical toolkit, and to provide enough background knowledge about them to give the researcher confidence when deciding to employ one (typically via a standard statistical computer package) in any practical application.

Historically, it was the burst of activity between 1795 and 1810 on the part of the aforementioned mathematicians Carl Friedrich Gauss, Pierre-Simon Laplace and Adrien Marie Legendre that heralded the beginning of statistical science as we know it. Indeed, it was more than just the beginning; between them, these three individuals provided the necessary framework and central methodology on which many of the statistical procedures developed in the next two centuries were based.

The starting point was not really statistical, but rather based in approximation theory. Among other interests, Legendre was working at the time on geodesy and astronomy, in particular on planetary orbits, and in both subjects practical measurement involved a certain amount of error. The combination of theory and observation usually led to a set of simultaneous linear equations expressing the errors in terms of a set of unknowns, and the problem was to find values of the unknowns that somehow made these errors as small as possible. One approach would be to set the errors all to zero and solve the equations, but the number of equations generally did not equal the number of unknowns so there was no unique solution. Legendre described the problem in 1795, and wondered how to balance the different errors in order to achieve an 'optimal' solution for the unknowns. As a simple example of this type of problem, consider weighing two objects on a faulty balance. Suppose weighing the two objects separately gives weights of 27 and 35 g respectively, while weighing them together gives 60 g. Clearly there are errors in the weighings, as these results are inconsistent. If we write a and b for the true weights of the two objects, then the three errors are $27 - a$, $35 - b$, and $60 - a - b$. If we set these three errors to zero, the resulting three equations in the two unknowns are inconsistent and do not have an exact solution; many different possible 'solutions' exist, each leading to different sets of three errors. Which is 'best'? The answer, which came to Legendre by way of inspiration and which he published in 1805, was that the best solution should be the one that minimised the sum of squared errors, and he set out various properties of this solution that convinced him of its optimality. In our example of the three weighings, we need to find values of a and b that minimise $(27 - a)^2 + (35 - b)^2 + (60 - a - b)^2$. This can be done very easily using differential calculus, and we find the required values to be $a = 26.33$ and $b = 34.33$, with a combined error sum of squares of 1.33.

Thus, the principle of least squares was established as a method for achieving a compromise, or balance, between sets of inconsistent measurements. This was taken forward by Gauss, who was the first person to apply statistical ideas via probability theory by looking for an appropriate 'probability function' for the errors that would justify the use of least squares for the estimation of the unknown quantities. He posited the existence of a set of unknown linear functions of a set of unknown parameters, with measurement of each function given with an error that came from this underlying probability function and interest focusing on estimation of the parameters. By the use of a heuristic argument allied to Bayes' theorem and uniform priors he was effectively led to an early version of the (much later) maximum likelihood principle, and then to propose what is now the very familiar standard normal function as the appropriate error function; for a long time, therefore, this was known as the Gaussian probability density function. This advance led to the possibility of statistical models being proposed for the 'explanation' of values of one variable (the 'response' variable) by the values of one or more other, 'explanatory', variables. In a typical model, the 'response' measurement could be viewed as made-up of a systematic part involving the combination of one or more unknown parameters with the 'explanatory' measurements, plus an error drawn from the normal distribution. The theory of least squares could be applied in order to estimate the unknown parameters, and normal theory inference would provide confidence intervals for them or tests of hypotheses about them.

Laplace provided the final piece of the jigsaw with his central limit theorem, already mentioned in the course of our discussion of statistical inference, which justified the assumption of normal errors for many practical statistical investigations. Laplace had derived this theorem in generality and without clarity regarding the final form of density function, but now he could adopt Gauss's function (and, incidentally, thereby provide a more solid rationale for it). Thus the way was now open for development of regression methods, both simple and multiple, in the later part of the nineteenth and early part of the twentieth centuries, and then the subsequent evolution of analysis of variance for designed experiments in the first half of the twentieth century. This body of techniques covered the requirements for analysis of the bulk of observational and experimental studies, and so formed the bedrock of statistical methodology in practice. We focus on these techniques in the remainder of this chapter, but first we must briefly consider association between measurements.

Correlation

Researchers in many different fields often measure several different quantities on each individual in a sample, and are then interested in deciding whether there is any relationship between these measurements. We will henceforward refer to a quantity being measured as a *variable*, in order to minimise mixtures of terms. If a relationship can be established between two variables, then they are said to be *associated, correlated* or *dependent*. Conversely, if no relationship exists between them then they are correspondingly described as *unassociated, uncorrelated* or *independent*. To analyse the association between two variables we must first decide on a way of measuring it. This will depend on the type of variables in question, that is, whether they are qualitative or quantitative. We will consider qualitative variables in a later chapter, so here focus just on quantitative ones.

If we want to investigate the association between two quantitative variables, the raw data will consist of pairs of values (one value for each variable) on a set of individuals. Some examples are as follows.

1. The rate of growth of a particular organism (variable y) is thought to depend in some way on the ambient temperature (°C, variable x). Ten separate experiments were conducted, giving the following results:

Temperature (x)	6	9	11	14	17	22	26	29	32	34
Rate of growth (y)	5	13	15	21	20	24	19	16	10	7

2. It is suspected that the maximum heart rate an individual can reach during intensive exercise decreases with age. A study was conducted on 10 randomly selected people, in which their peak heart rates were recorded when performing exercise tests, with the following results:

Age (x)	30	38	41	38	29	39	46	41	42	24
Heart rate (y)	186	183	171	177	191	177	175	176	171	196

3. The coursework assessment (CA) and final year exam marks of 10 students in a statistics class are given below:

CA mark (x)	87	79	88	98	96	73	83	79	91	94
Final mark (y)	70	74	80	84	80	67	70	64	74	82

We can investigate association graphically by plotting a *scatter diagram* for such a pair of variables. This is a simple plot in which each individual is represented by a point whose coordinate on the horizontal axis is given by that individual's value on one of the variables (usually the x variable) and whose coordinate on the vertical axis is given by its value on the other (y) variable. The scatter diagrams for the three examples above are shown in Figure 6.1 (a)–(c). The first pair of variables shows a *curvilinear* association (one variable rises then falls as the other rises); the second pair shows a *negative linear* association (one variable falls approximately linearly as the other rises); while the third pair shows a *positive linear* association (the variables fall or rise together, approximately linearly).

We saw in Chapter 2 that the variance of a set of n measurements is the slightly adjusted average of the squared differences between each measurement and the mean of those n measurements (the slight adjustment being division by $n - 1$ instead of by n). So in each of the examples above we can calculate the variance of the x measurements and the variance of the y measurements. Taking the square root of each variance gives us the standard deviation of those measurements; let us denote the standard deviations of the x and y measurements in a particular case by s_x and s_y, respectively. Now in each case, we can additionally compute the *covariance* between x and y, which we denote by s_{xy}. This is done by first finding the differences between each x value and the x mean and the differences between each y value and the y mean, and then calculating the similarly slightly adjusted (i.e. dividing by $n - 1$) average of the products of these pairs of differences across individuals. The variances measure the amount of variation in each of x and y separately, while the covariance measures the amount by which they vary jointly. But the actual value of s_{xy} in any particular example will depend on the scale of values of x and y: for example, in the three cases above we have respectively -1.11, -53.93 and 46.78 for s_{xy}. So it is not possible to infer the strength of relationship between x and y just from s_{xy}; we must first standardise it onto a common scale to make it consistent across different data sets, and this is done by dividing it by the product $s_x s_y$ of the two standard deviations. The result,

$$r_{xy} = \frac{s_{xy}}{s_x s_y}$$

is known as the *correlation coefficient* between x and y.

The correlation coefficient is a summary measure of the strength of a *linear* association which, broadly speaking, measures the closeness of the

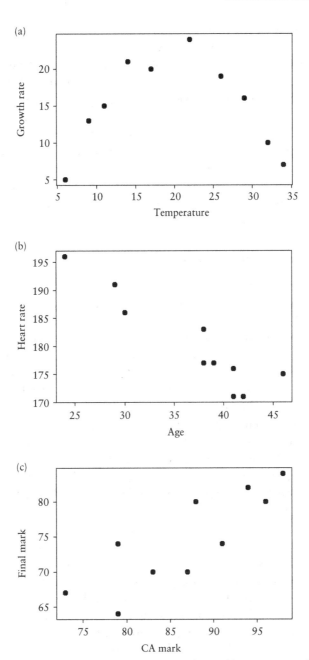

Figure 6.1. Scatter diagrams for three pairs of variables. (a) Growth rate versus temperature; (b) heart rate versus age; (c) final mark versus CA mark.

points in a scatter diagram to a straight line. It will have value between $+1$ and -1: negative values signify negative association between the variables, and the closer the value is to plus or minus 1, the closer are the points to a straight line. A value equal to plus or minus 1 indicates perfect linear relationship, that is, the points lie exactly on a straight line, while a value close to zero signifies no discernible linear association. The latter most commonly arises if the points form a dispersed cloud with no indication of a linear relationship, but a correlation close to zero can arise also when there is a curvilinear relationship between the variables. (Note also that a value close to plus or minus 1 can arise when there is so-called 'spurious correlation': all points except one form a dispersed cloud of zero correlation, but the remaining point is far removed from this cloud and creates an impression of linear association.) For the three examples above we have values -0.177, -0.929 and 0.839, respectively, for r_{xy}, which accord with the above interpretations in the light of the scatter diagrams in Figure 6.1.

Explanation and prediction

While assessment of the strength of (linear) relationship between two variables is obviously very useful, it may not go far enough in many instances. If there is some relationship between the variables then a precise description of this relationship should be of great value. It may, for example, give some insight into the mechanism governing the behaviour of some physical situation and hence provide an explanation for it. Alternatively, circumstances are frequently encountered in which one variable is costly or difficult to measure but the other is not. Knowledge of the relationship between the two variables makes it possible for us just to measure the 'easy' one and then to 'predict' the corresponding values of the 'difficult' variable much more accurately than by guesswork. Thus cost and labour can be kept to a minimum, while ensuring reasonable accuracy of results. In general, knowledge of such relationships can be very useful, whether in analysis, decision-making or planning some future course of action.

We are therefore often interested in the influence of one variable on another. However, there is one fundamental difference here from simply measuring the association between two variables. The correlation coefficient is a symmetric quantity: it does not matter which variable is labelled x and which is labelled y, as the value of the correlation coefficient will be the same whichever labelling is used. Now, however, it is important to identify which is the 'dependent' variable and which is the 'explanatory' one, that is, which is the variable that we are predicting and which is the one forming the basis of the prediction. It is conventional to denote the former by y and the latter by x. The prediction is effected by means of a *regression analysis* in which we *regress y on x*.

Usually it is clear which way round the variables ought to be. For example, we would typically wish to predict heart rate from age and final mark from CA mark in examples (2) and (3) above; the reverse predictions would not really make sense (so the variables have already been labelled accordingly). Sometimes, however, either prediction could be reasonably contemplated and in these circumstances we must consider carefully which way round the analysis should be conducted. Results of a regression of y on x cannot in general be deduced from those of a regression of x on y, and two separate analyses have to be conducted if we want to investigate both possibilities.

Simple linear regression

We start with the simplest case of a linear relationship between two variables. The analysis of such relationships is relatively straightforward and they arise surprisingly often in practice. Indeed, a linear relationship is often a perfectly adequate approximation to a more sophisticated relationship, for all practical purposes.

Suppose first that there is a perfect linear relationship between the variables x and y. This means that all points on the scatter diagram will lie on a straight line, as shown in Figure 6.2. How can we describe such a relationship?

A perfect linear relationship means that every change of 1 unit in x results in a constant absolute change in y. Let this change in y be denoted by b, and suppose the line crosses the y-axis at a. Then: when $x = 0$, $y = a$; when $x = 1$, $y = a + b$; when $x = 2$, $y = a + 2b$; and so on. Hence at a general value g of x, $y = a + bg$. This is the *equation* of a straight line that has intercept a and *slope* b. Any straight line connecting variables x and y can therefore be expressed as the equation $y = a + bx$ for the appropriate values of a and b. For the line shown in Figure 6.2, at $x = 0$ we have $y = 1.5$ and for every unit increase in x, the y value increases by 2. Hence $a = 1.5$ and $b = 2$, so the equation of the line is $y = 1.5 + 2x$. Using this equation,

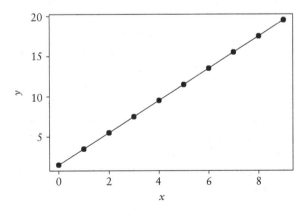

Figure 6.2. Scatterplot for a perfect linear relationship between x and y.

we can predict the value of y for any specified value of x, for example, at $x = 10$ we have $y = 1.5 + 2 \times 10 = 21.5$.

Of course, the scatter diagram for any real set of data will not have points lying exactly on a straight line, so the purpose of regression analysis is to find the *best* straight line that can be fitted to the points. This is where the principle of least squares comes in. Suppose we have a candidate line with equation $y = a + bx$ for some specified a and b. Corresponding to every value of x in our sample, we can find the value, z say, on the line. But we also have the sample y value corresponding to this x value, and, since we are predicting y from x, the difference between this y value and z is the discrepancy at that x value. The principle of least squares tells us to find that line for which the sum of squared discrepancies over the whole sample is least, and this is the best fitting line as it finds the best 'balance' between all the discrepancies. It is worth noting that this is in fact what most of us do intuitively when we fit a straight line to a set of points 'by eye'.

Naturally, we do not have to try all sorts of possible lines laboriously in order to find the best one, because the problem is solved by the use of differential calculus and some algebraic manipulation. The best fitting line has slope $\hat{b}$ given by the covariance between x and y divided by the variance of x, and intercept $\hat{a}$ given by the mean of y minus $\hat{b}$ times the mean of x. Thus, in terms of the previously defined quantities, we have

$$\hat{b} = \frac{s_{xy}}{s_x^2} \quad \text{and} \quad \hat{a} = \bar{y} - \hat{b}\bar{x}.$$

Moreover, given the earlier definition of the correlation coefficient r_{xy} between x and y, we have the equivalent expression for the slope:

$$\hat{b} = \frac{r_{xy} s_y}{s_x}.$$

Most standard computer packages will obtain these values directly from the raw x, y values. Note that we write 'hats' above both a and b to emphasise that these are just estimates obtained from the data, and not necessarily 'true' values.

There is some standard terminology associated with this technique. The line $y = \hat{a} + \hat{b}x$ is called the (estimated) regression line of y on x. It always passes through the point corresponding to the means of x and y. The slope $\hat{b}$ gives the *average* increase in y values for a unit increase in x (or the *expected* increase when predicting values). It is also known as the *regression coefficient*, and the intercept $\hat{a}$ is also known as the *regression constant*. Consider any particular pair of values $x = c$, $y = d$ in the data set . The value on the line at $x = c$ can be denoted by $\hat{y}$ and is given by $\hat{y} = \hat{a} + \hat{b}c$. It is called the *fitted value* at $x = c$, and the discrepancy $e = d - \hat{y}$ between the observed and fitted values at $x = c$ is called the *residual* at $x = c$.

Let us pause briefly at this juncture to consider why the rather unnatural term 'regression' is used for the fitting of lines through sets of points. The following explanation was provided by J.M. Bland and D.G. Altman in the *British Medical Journal* (1994). It appears that the term was first used in the article 'Regression towards mediocrity in hereditary stature', published in the *Journal of the Anthropological Institute* (1886) by Francis Galton, who studied heights of children in relation to the average (i.e. mid-) heights of their parents. Overall, children and parents had the same mean heights, but Galton noticed that if he restricted attention to just those parents having a particular mid-height then the mean height of their children was closer to the mean height of all children than this mid-height was to the mean height of all parents. Similarly for children of a particular height, the mean mid-height of their parents was closer to the mean height of all parents than their height was to the mean height of all children. This 'regression (i.e. going back) towards mediocrity' subsequently became known as 'regression towards the mean'. Moreover, taking each group of parents by mid-height, calculating the mean height of their children and plotting the resulting scatter diagram revealed that the points lay close to a straight line. This therefore became known as the regression line, and the whole process of fitting the line became known as regression.

Furthermore, consideration of the second expression above and subsequent interpretation for the slope $\hat{b}$ of the line gives a simple explanation for such a regression towards the mean. A change of one standard deviation in x will on average produce a change of r_{xy} standard deviations in y, and unless x and y have a perfect linear relationship this will therefore be less than one standard deviation change in y. Thus in virtually all practical situations (since the case $r_{xy} = 1$ is extremely rare), for a given value of x the predicted value of y will be closer to its mean than x is to its mean.

Returning now to the general development, the correlation coefficient r_{xy} between x and y can also be used to measure how well the regression line fits the data. Recollect that the closer the value of r_{xy} is to either $+1$ or -1, the closer do the points lie to a straight line. Since we are not worried about the *sign* of the coefficient, it is preferable to calculate the square of the correlation, that is, r_{xy}^2, and thereby to remove negative signs. Moreover, as we shall see below, r_{xy}^2 also measures the proportion of variability in y that is 'explained' through the relationship between y and x, which is effectively a measure of fit of line to data. So r_{xy}^2 lies between 0 and 1; the value 1 indicates a perfect fit of points to straight line, a value between 0.8 and 1.0 would generally be regarded as a very good fit, while values between 0.5 and 0.8 might variously be deemed adequate fits depending on subject area—the higher values being necessary in science, but the lower ones being tolerable in the social sciences (where there is large between-individual variability). However, values lower than 0.5 would indicate that the line does not really fit the data.

For the age/heart rate data, the fitted regression line is $y = 222.25 - 1.14x$ and the measure of fit is $r_{xy}^2 = 0.863$, while for the CA/exam marks we have the fitted line $y = 14.35 + 0.693x$ and a measure of fit of $r_{xy}^2 = 0.704$. A final point to note here is that r_{xy}^2 is equivalently the squared correlation coefficient between each observed y value and its corresponding fitted value $\hat{y}$, which is a useful equivalence when considering more complicated regression models.

Testing significance of regression

Consider the variability exhibited by the dependent variable y. This is measured by the variance of y, or equivalently (just ignoring the constant factor $n - 1$) by the sum of squared discrepancies between the y values and their mean. However, if there is a linear relationship between the explanatory variable x and the dependent variable y, then some or all of the variability in y is 'explained' by this relationship (in the sense that variation in x values will perforce induce variation in y values). In the scatter diagram of Figure 6.2, the perfect linear relationship between y and x means that *all* the variability in y values has been induced by the variability in x values. In the case of the age/heart rate and the CA/exam scatter diagrams of Figure 6.1, however, only *some* of the y variability is so explainable, because here there is additional variation over and above the linear relationship (as the fit is not perfect in these instances).

In fact, it is possible to calculate exactly how much of the y variation is caused by the linear relationship and how much is attributable to 'noise' or random fluctuation. Consider Figure 6.3, in which the fitted regression line is superimposed on the third scatter diagram of Figure 6.1 for the CA/exam marks.

The mean of the final marks is 74.5, and it is easy to see that the vertical distance between any point in the diagram and 74.5 is made up of two parts: the distance between the point and the regression line and the distance between the regression line and 74.5 (either of which can be

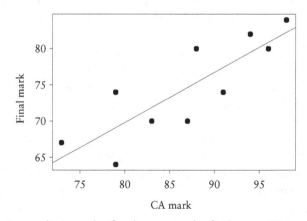

Figure 6.3. Scatter diagram plus fitted regression for final versus CA marks.

positive or negative). The latter distance gives the contribution of the linear relationship to the discrepancy between that point and 74.5, while the former gives the contribution of the 'lack of fit' of that point to the line. It can be shown mathematically that the sum of squares of all deviations of points about the mean, the 'total sum of squares' (TSS), is equal to the sum of squares of these two sets of distances: the sum of squares of the second set of distances (the contributions of the linear relationship) is known as the 'regression sum of squares' (RSS) and the sum of squares of the first set of distances (the contributions of the lack of fit) is known as the 'residual sum of squares' or the 'error sum of squares' (ESS). Moreover, it can be shown also that

$$\text{RSS} = \text{TSS} \times r_{xy}^2 \quad \text{and} \quad \text{ESS} = \text{TSS} \times (1 - r_{xy}^2).$$

The relative sizes of these two sums of squares will generally be of interest in any regression analysis, as the greater the RSS, or the smaller the ESS, the more variation is explained by the regression. It is therefore convenient to summarise this breakdown of variation in a table known as the *Analysis of Variance* or ANOVA table.

Source	Sum of squares	Degrees of freedom	Mean squares	F ratio
Regression	RSS	1	RMS = RSS	RMS/EMS
Residual	ESS	$n - 2$	EMS = ESS/$(n - 2)$	
Total	TSS	$n - 1$		

We first note that the ratio RSS/TSS gives the value of r_{xy}^2, which justifies our previous reference to this quantity as the proportion of variability in y that is 'explained' through the relationship between y and x. So if the table has been produced by a statistical package such as SPSS, SAS or Minitab, for example, then one use of it might be for the calculation of r_{xy}^2. However, a more fundamental use is for testing whether there is evidence of a worthwhile relationship between x and y. But such a test moves us from simple fitting of a straight line through points to the testing of a hypothesis about a putative population. Although the mechanics are still based round least squares, the underlying assumptions have altered significantly—in effect we have moved from Legendre's justification of the method to that of Gauss, and we now need to incorporate a normal distribution into the process. This is done formally by assuming that each y value is equal to its prediction from the linear model plus an 'error', and these 'errors' are independent observations from a normal distribution having mean zero. So we are now in effect seeking to minimise a set of normal 'errors' or 'departures' from a postulated linear model. To see how the testing is carried out, however, we first need to explain the entries in the ANOVA table.

Associated with each sum of squares is a number of *degrees of freedom*. We have already seen in Chapter 4 that there are $n - 1$ degrees of freedom when estimating the population variance from a sample and the same holds for the sum of squared deviations in the present case also, so this value is entered next to the TSS. There is just one explanatory variable in simple regression, so there is 1 degree of freedom for RSS. Finally, to ensure that the numbers in the column add up correctly, there are $n - 2$ degrees of freedom for the residual sum of squares ESS. An alternative (perhaps more intuitive) justification is that the number of degrees of freedom for ESS equals the number of observations (n) less the number of model parameters that have been estimated (intercept a and slope b of the line).

The entries in the 'mean squares' column are obtained on dividing the entries in the 'sum of squares' column by the corresponding entries in the 'degrees of freedom' column, so that the regression mean square RMS is equal to RSS while the residual mean square EMS is ESS/$(n - 2)$. The final F ratio entry is obtained by dividing the regression mean square by the residual mean square. We then use the ANOVA table to deduce the 'significance' of the fitted regression by considering the value in the F ratio column. A large value here (as determined by reference to F tables) is indicative of a significant regression, that is, the explanatory variable x has an important (linear) relationship with y, whereas a small value suggests that there is no real (linear) relationship between x and y. If y does not depend on x then that is the same as saying that the true regression slope b is zero. Thus the ANOVA significance test is equivalent to a test of the null hypothesis that b is zero against the alternative that it is non-zero. A computer package such as those mentioned above will print out the significance level p of any computed F ratio value, and will also provide the equivalent t-statistic for testing b (but remember that *small p* values indicate 'significance' while large ones indicate 'lack of relationship'). Alongside the test for $b = 0$, the package will also usually supply either a confidence interval for b or a standard error from which the confidence interval is easily calculated.

To illustrate these calculations, consider the final mark versus CA mark regression. The sum of squared discrepancies of the $n = 10$ y values from their mean, TSS, is 414.5 and $r_{xy}^2 = 0.703$, so these values enable the table to be filled out as follows:

Source	Sum of squares	Degrees of freedom	Mean squares	F ratio
Regression	291.81	1	291.81	19.03
Residual	122.69	8	15.336	
Total	414.50	9		

The computed F ratio is large for these degrees of freedom, and reference to tables of the F distribution gives the significance level as less than 0.01, so we conclude that there is indeed a significant linear relationship between CA mark and final mark.

Multiple regression

In a simple linear regression, it is assumed that the responses for a dependent variable y can be 'explained' mainly by those of one other variable. Such an assumption implies that the multitude of other factors, which may influence the dependent variable, are only of minor importance. This may often be invalid, and consequently we are led to studying the dependence of y on a *set* of explanatory variables, say $x_1, x_2, \ldots, x_k$. For example, we may conjecture that the rent charged for a commercial property in London (y) depends on a number of factors such as its floor space (x_1), its ceiling height (x_2), its distance from the nearest motorway (x_3), etc., and we can obtain values for each of these variables on a number of properties. One way to explore this hypothesis is to try and fit from our data a simple relationship between y and all the x variables. If the fit seems adequate then we can use this relationship in future decisions, perhaps predicting a value of y for given values of the other variables. Alternatively, examination of the relationship may give us some insight into the mechanism governing such a set of data and help us with some future planning (e.g. in the specification of a warehouse to be constructed).

Extension of the earlier ideas suggests that the simplest relationship connecting y and a set of x variables is a relationship having the general form

$$y = a + bx_1 + cx_2 + \cdots + dx_k.$$

In this equation, a, b, c, d, etc. are *parameters*, and we can again use the method of least squares to find the 'best' estimates $\hat{a}, \hat{b}, \hat{c}, \ldots, \hat{d}$ for a particular set of data. These are the values that minimise the sum of squared discrepancies between each observed y value and its 'fitted value' $\hat{y}$ where

$$\hat{y} = \hat{a} + \hat{b}x_1 + \hat{c}x_2 + \cdots + \hat{d}x_k.$$

When we fit such a model we say that we are conducting a *multiple regression* of y on $x_1, x_2, \ldots, x_k$.

The mathematics needed to establish the form of these estimates is considerably more sophisticated than that for simple regression above, and indeed it is not possible to write this form down in simple fashion. However, presenting a package such as SPSS, SAS or Minitab with the relevant set of data and asking for the estimates to be computed is a trivial task that is typically performed in a very short time. This produces numerical estimates of the regression parameters.

Notice that the multiple regression model above is quite flexible, and can be adapted to investigate quite a range of relationships. For example, the relationship between rate of growth of an organism y and ambient temperature x earlier in this chapter would not be fitted well by a simple regression because the scatter diagram in Figure 6.1 shows a curve rather than a straight line. However, if we put the *squares* of the values of x into a new variable z, then a multiple regression of y on x and z will fit a quadratic relationship between y and x. This example illustrates a frequent source of confusion about the term 'linear model'. Users who are aware that multiple regression procedures are only applicable to linear models sometimes hold back from fitting quadratic or higher order polynomials to data, in the mistaken belief that such models cannot be fitted by multiple regression. In fact, when we speak of a linear model we mean 'linear in the parameters', and not 'linear in the data'. By 'linear in the parameters' we mean that a unit change in any parameter value will produce the same change in the response variable, *whatever the actual value of the parameter*. For example, if the coefficient of x_1 in $y = a + bx_1 + cx_2 + \cdots + dx_k$ changes from b to $b + 1$, then the change in y is $a + (b + 1)x_1 + cx_2 + \cdots + dx_k - (a + bx_1 + cx_2 + \cdots + dx_k)$, which is x_1 whatever the actual value of b. This is true for all the parameters of the multiple regression model, and thus verifies that it is a linear model. Expressing the squares of x as a new variable z shows that the linearity in the parameters is maintained. We will look at non-linear models in the next chapter.

To measure how well the multiple regression model fits a set of data, we use the equivalence established earlier for simple regression, and calculate the squared correlation between the observed y values and their corresponding $\hat{y}$, that is, their 'fitted', values. This squared correlation is usually denoted R^2, to distinguish it from the simple regression case where there is only one x measurement, and is often referred to as the 'coefficient of determination'. Comments given above about values of r_{xy}^2 in relation to goodness of fit for simple regression hold equally for R^2 and multiple regression. Moreover, significance of the relationship between y and $x_1, x_2, \ldots, x_k$ can again be tested via ANOVA. The form of the ANOVA table in this case is only slightly different from before:

Source	Sum of squares	Degrees of freedom	Mean squares	F ratio
Regression	RSS	k	RMS = RSS/k	RMS/EMS
Residual	ESS	$n - k - 1$	EMS = ESS/$(n - k - 1)$	
Total	TSS	$n - 1$		

The main difference stems from the 'degrees of freedom'; since there are k explanatory variables the regression sum of squares RSS, has k degrees of freedom. This has a knock-on effect on ESS, which therefore has $n - k - 1$ degrees of freedom (justified either because the sum of degrees of freedom is $n - 1$, or because $k + 1$ parameters have been estimated from n sample observations) and on the RMS, which is now RSS divided by k on following the usual rules. Of course, the actual calculation of RSS and ESS differs from before, and because of the sophisticated mathematics involved in multiple regression it is not possible to give details here. However, the interpretations are the same: RSS is the part of the variability of y that is attributable to its linear relationship with $x_1, x_2, \ldots, x_k$, while ESS is the part of the variability that is unexplained. All statistical packages effect the necessary computations very quickly.

All other features of the ANOVA table and its interpretation follow the same lines as before. The ratio RSS/TSS again yields the value of R^2 for the regression, and the value in the F ratio column enables the significance of the fitted regression to be tested as before. This is now an 'omnibus' test on the regression coefficients, the null hypothesis being that *all* the coefficients are zero (so y is not dependent on any of the $x_1, x_2, \ldots, x_k$) while the alternative is that *at least one* coefficient is non-zero. Computer packages will print out the significance level p of the F value in the usual way. However, it should never be forgotten that in order to conduct significance tests we have to make distributional assumptions, and as for simple regression we here assume that each y value is equal to its prediction from the linear model plus an 'error', and these 'errors' are independent observations from a normal distribution having mean zero.

Model building

A situation commonly encountered is when a researcher has observed a response variable y and a set of explanatory variables $x_1, x_2, \ldots, x_k$, has fitted a multiple regression and found it significant, but has now collected a further set of q explanatory variables $x_{k+1}, x_{k+2}, \ldots, x_{k+q}$ and wishes to know whether the model will be improved by adding this set of explanatory variables to the former set. The first thing to note is that the apparent fit of a model will *always* be improved by adding extra explanatory variables to it, but the addition is only warranted if the improvement is 'significant'. If it is not significant, it means that the apparent 'improvement' is simply commensurate with the noise in the system, and the larger model is inefficient by comparison with the smaller one. To measure whether the extra variables are significant or not, we extend the previous ANOVA table in the following way. We first fit the regression including all $k + q$

explanatory variables, and form the usual ANOVA table:

Source	Sum of squares	Degrees of freedom	Mean squares	F ratio
Regression (all $k + q$)	RSS	$k + q$	RMS = RSS/$(k + q)$	RMS/EMS
Residual	ESS	$n - k - q - 1$	EMS = ESS/$(n - k - q - 1)$	
Total	TSS	$n - 1$		

We then fit the regression for just the first k variables $x_1, x_2, \ldots, x_k$, and denote the regression sum of squares by RSSA. The difference RSS − RSSA will always be positive, since by the above argument the larger model is always 'better' so has a larger regression sum of squares. This difference, known as the *extra sum of squares* due to $x_{k+1}, x_{k+2}, \ldots, x_{k+q}$, is attributed to the effect of the q extra explanatory variables, *in the presence of the first k explanatory variables*, and it has q degrees of freedom. The relevant quantities can be incorporated into our ANOVA table by subdividing the 'due to regression' line into two lines as follows:

Source	Sum of squares	Degrees of freedom	Mean squares	F ratio
Regression (all $k + q$)	RSS	$k + q$	RMS = RSS/$(k + q)$	RMS/EMS
Regression (first k)	RSSA	k	RMSA = RSSA/k	RMSA/EMS
Regression (extra q)	RSS − RSSA	q	RMSE = (RSS − RSSA)/q	RMSE/EMS
Residual	ESS	$n - k - q - 1$	EMS = ESS/$(n - k - q - 1)$	
Total	TSS	$n - 1$		

This table thus gives rise to three possible tests: the significance of the model that includes just the first k variables is tested by referring the calculated value of RMSA/EMS to F tables; the significance of the model that includes all $k + q$ variables is tested by referring the calculated value of RMS/EMS to F tables; and the significance of the q extra variables over and above the k original ones is tested by referring the calculated value of RMSE/EMS to F tables. In all cases, reference to F tables will provide the significance level p of the corresponding test. But it should be stressed that, of course, the test of extra explanatory power of the added q variables depends crucially on which other variables are already present in the model. In particular, the regression sum of squares for the extra q variables over and above the original k variables, RSS $-$ RSSA, is generally *not* the same as the regression sum of squares (RSSB, say) when the model contains just these q variables. If these two sums of squares *do* turn out to be equal, then RSS $-$ RSSA = RSSB, so that RSS = RSSA + RSSB. In such a case, the regression sum of squares when both sets of variables are present is a simple addition of the two regression sums of squares when each set is present by itself. In this case we say that the two sets of variables are *orthogonal*, and to assess the effect of either set it is immaterial whether the other is present or not. Orthogonal sets of variables are relatively uncommon in regression, but we will see later that they readily occur in other situations.

Many regression studies are exploratory, in that we are looking for a 'good' set of explanatory variables with which to predict a dependent variable, often from a large pool of potential explanatory variables. Thus we are frequently led to a systematic search through the pool of potential variables, in the hope that we will find a small number of them that produce a good fit when a multiple regression of the dependent variable on them is conducted. This process has a number of pitfalls for the unwary, so the following points may be useful to bear in mind when considering such a study.

The R^2 value will *always increase* when an extra explanatory variable is added to an existing set of explanatory variables, for the reasons given above. So if this statistic is being used as an indicator of whether or not to include a particular variable, that variable will only be worth retaining if the increase is substantial. To eliminate the need for making such a subjective decision, many computer packages also provide a quantity called the *adjusted R^2* in the output of a multiple regression analysis. If there are k explanatory variables in the model, the adjustment consists of multiplying R^2 by $(n-1)/(n-k-1)$ and subtracting $k/(n-k-1)$ from the result. The adjusted R^2 compensates for sample size and degrees of freedom in the assessment of worth of added variables, so does not necessarily increase when extra explanatory variables are introduced into the model. Any increase in this statistic may therefore be taken as indicative of explanatory worth of these variables.

Most computer packages, when printing out the regression coefficient estimates, will also print out alongside each estimate its standard error plus a t-value that can be referred to student's t-distribution. The standard error can be used to find a confidence interval for the coefficient, while the t-value tests the null hypothesis that the true coefficient is zero (i.e. the corresponding variable contributes nothing) against the alternative that it is not zero (i.e. the corresponding variable contributes something). However, it should always be remembered that these tests are only valid *when all the other variables are present in the regression*, and the conditional nature of the relationship being assessed is actually part of the hypothesis. Thus a significant coefficient indicates that it is worthwhile adding the given variable to the others in the set, and conversely a nonsignificant one suggests that that variable can be removed from the set. But once an addition or removal has been made the regression needs to be recomputed because the various coefficient estimates and t-values *will now change*, as there are different variables in the set (unless all the variables are orthogonal to each other, a highly implausible situation in practical regression studies). The (often employed) practice of deleting simultaneously all those variables that do not have significant coefficients is not correct, as deleting some of them might convert the coefficients of others to significance (or, conversely, adding some significant variables may reduce previously significant coefficients to nonsignificance).

It is thus clear from the foregoing that full exploration requires a possibly lengthy sequence of trials of different multiple regressions, perhaps involving scrutiny of all possible 2^k models formed from k explanatory variables. If k is at all large, this is a formidable computing task so most computer packages have a command that will initiate a systematic search of a much-reduced set of possibilities automatically.

This command can usually be found under the title 'selection of variables', and various strategies for reducing the effort of the search include 'forward selection' (adding significant variables one by one from the null model until further addition becomes nonsignificant), 'backward elimination' (leaving out nonsignificant variables one by one from the full set until further deletion becomes significant) or 'stepwise search' (a judicious blend of the previous two strategies). The last-named is usually the most comprehensive, as it allows elimination of variables as well as addition of variables at each stage so caters for the possibility of some variables changing their significance on subsequent addition or deletion of other variables. These strategies are fully automatic, usually only requiring the user to specify at the outset either the significance level or the effect size (i.e. the F ratio value) to use as threshold for inclusion or deletion of a variable at each stage. They lead to fast computation so have the ability to deal with large numbers of variables, but will not necessarily produce the 'best' model at the end of the process. However, the final model is usually close to if not actually 'best', so this is the strategy that is most commonly employed in practical model building.

One final cautionary note should be sounded in the context of stepwise searching for optimal models. When sample sizes are very large (an increasingly common situation given present-day computer storage capabilities), very small departures from a null hypothesis are often sufficient to trigger a significant result in any hypothesis test. This should be borne in mind whatever the context of the test, and care must be taken to distinguish between 'statistical significance' and 'practical significance' in any given situation. This point is just as pertinent in stepwise model selection, where the danger is that too many explanatory measurements may be deemed to be significant, and hence the final model may be far too complicated for practical use. To mitigate this danger we need to use either a more stringent significance level or a larger effect size as the threshold for inclusion of terms in the model.

Analysis of variance

We have seen that the analysis of variance is a common device for partitioning the sum of squares in regression analysis, and hence in providing significance tests for various hypotheses of interest. However, its formal introduction into common statistical usage came in a different context, and was another of Fisher's many contributions to the subject. Indeed, it was probably the earliest and the most far-reaching of his practical innovations, dating from a paper he published in 1921, but it could be said that that all these developments were somewhat fortuitous in their genesis. John Lawes had started agricultural experimentation in the 1840s at Rothamsted Experimental Station in the United Kingdom, and continued his experiments with Joseph Gilbert and a few co-workers into the twentieth century. However, many of the results and data lay there unanalysed until the appointment in 1919 of Fisher, hitherto a teacher of mathematics at various schools. His later-expressed and somewhat disparaging view was that he had been appointed to 'rake over the old muck-heap', but we can all be thankful that he did far more than just that! Examining such a mass of data inspired a more scientific approach, and many consider Rothamsted to be the birthplace of modern statistical theory and practice. The analysis of variance was one of the earliest tools Fisher developed, and it has been a cornerstone of statistical analysis ever since.

To see how it works, let us consider some details in the simplest situation to which it is applicable, namely the testing of differences between means of a set of populations from each of which we have sample data. The case of just two populations can be handled by way of a two-sample t-test, so the analysis of variance is usually only brought into play when there are more than two populations. An example of such a situation is the set of data shown in Table 6.1. This came from an experiment to investigate the rate of wear of tyres, and whether this rate differed between the four positions that they can occupy on a car. The tyres were fitted to a car, the car was driven at a fixed speed for a fixed distance and the reduction in depth of tread was measured (in hundredths of a millimetre) for each tyre. The process was repeated nine times with a different set of tyres each time. Thus the data can be considered as samples of size nine from each of four populations, namely the populations of tyre wear for front offside, front nearside, rear offside and rear nearside car positions, and the objective of the analysis is to determine whether these four populations all have the same mean ('car position does not affect mean tyre wear') or whether at least one mean differs from the others ('car position does affect mean tyre wear').

Table 6.1. Data from experiment on tyre wear

Car position	Reduction in depth of tread		
Front offside (FO)	20.935	17.123	29.590
	19.013	15.919	28.092
	20.332	15.285	28.304
Front nearside (FN)	18.279	14.815	19.973
	21.200	11.280	20.096
	19.389	12.153	20.477
Rear offside (RO)	28.535	37.227	30.529
	27.998	38.853	29.177
	30.073	40.017	30.795
Rear nearside (RN)	20.182	34.340	29.023
	18.792	34.707	28.176
	19.203	36.307	28.701

For these data, the following calculations can be easily verified:

(1) the means of the nine values in the four positions are 21.62 (FO), 17.52 (FN), 32.58 (RO) and 27.71 (RN);

(2) the mean of all the 36 values is 24.86;

(3) the sum of squares of all 36 values about 24.86 (i.e. the 'total' sum of squares, TSS) is 2112.13;

(4) the sums of squares of the nine values about their own means for the four positions are 251.86 (FO), 114.07 (FN), 178.87 (RO) and 378.32, and the sum of these four values (i.e. the 'within-group' sum of squares, WSS) is 923.12;

(5) so the difference in these two sums of squares is TSS − WSS = 2112.13 − 923.12 = 1189.01; this is usually termed the 'between-group' sum of squares, BSS.

Fisher's essential insight was to note that if all the population means are the same, then the sum of squared differences of all the observations from their overall mean (TSS) should be comparable to the aggregate of the sums of squared differences of the observations in the separate samples from their own means (WSS), because the individual sample means should all be close to the overall mean. However, if there is some variation among the population means, then WSS will not be affected (because each mean is first removed from the values in that group), but TSS will be inflated (because the variation in means will increase the variability of all the individual observations). So the discrepancy BSS between these two sums of squares measures the lack of equality of all the population means. If the discrepancy is 'small' then this would support the view that the population means are all equal, but a 'large' discrepancy would support the view that there is some difference among the population means.

So it just remains to determine whether the calculated BSS value is 'small' (in which case the evidence is in favour of equality of population means) or 'large' (in which case we would conclude that there is some difference between them). Fisher at first tackled this nonparametrically, using a permutation test. His argument was that if there is no difference between the population means then there should be no essential difference between the BSS values when the data are randomly permuted between the populations, but if there is a difference between the population means then such permutation will *reduce* the sample mean differences and so make the original BSS value greater than most of the 'permuted' BSS values. Thus, for the tyre data, if many such random permutations are conducted and the BSS value of 1189.01 is within the main bulk of the set of resulting BSS values then we would conclude that there is no difference among the population means, while if 1189.01 comes in the top $p\%$ of the BSS values then we would reject equality of population means at the $p\%$ level of significance.

Such a permutational test is an exact one that makes no distributional assumptions, but it requires much computing effort. Indeed, it taxed the capabilities of most workers in the 1920s on even small sets of data. So Fisher made the further assumption of normality of parent populations, and developed a parametric test based on the analysis of variance. The first step was to find algebraic expressions for BSS and WSS in general, and to establish the equality TSS = BSS + WSS mathematically. In fact, BSS is simply given by the sum of squared differences between each group mean and the overall mean, weighted by the number of observations in each group. If there are k populations under consideration and n observations altogether, then the $n - 1$ degrees of freedom for TSS are divided into $k - 1$ for BSS and $n - k - 1$ for WSS. Assuming a constant variance σ^2 in each population, WSS divided by $n - k - 1$ always gives an estimate of σ^2, while if there is no difference in the population means then BSS divided by $k - 1$ gives a second, independent, estimate of σ^2. Fisher had already studied the behaviour of the ratio of two independent estimates of the same variance in normal populations, and had derived the sampling distribution of this ratio (known originally as the variance ratio distribution but subsequently as the F distribution after Fisher). So referring the calculated ratio to this distribution will show whether it was consonant with the bulk of the distribution (implying that the population means were equal) or whether it was 'extreme' (implying some difference between the population means).

All the calculations can be written in an analysis of variance table similar to the ones we have seen in regression, but replacing the 'Regression' and 'Error' sums of squares by BSS and WSS respectively and using the values given above for the degrees of freedom. All subsequent calculations in this

Table 6.2. ANOVA table for tyre wear data

Source	Sum of squares	Degrees of freedom	Mean squares	F ratio
Between groups	1189.01	3	396.34	13.7
Within groups	923.12	32	28.85	
Total	2112.13	35		

table are then conducted as before. So for the tyre data we have the ANOVA table shown in Table 6.2.

With an assumption of normality for the data we thus refer to tables of the F distribution for the test of equality of means, and we now find that the value 13.7 is indeed 'large' at all the usual significance levels, so we conclude that there is evidence of some difference among the four population means. Moreover, since 28.85 is an estimate of the common population variance σ^2 we can now use the standard methods from Chapter 4 to construct a confidence interval for each population mean in order to explore this difference.

Although we started this section by saying that the case of two populations could be handled by a two-sample t-test, there is no bar to the use of analysis of variance in this case also and the same conclusions will be reached as in the two-sample t-test. In fact, it can be shown mathematically that the F ratio of the analysis of variance test equals the square of the t-statistic in the latter test, and exactly the same significance level is obtained from statistical tables in the two cases. Finally, it is also interesting to note that in these days of cheap computing power there is no longer any problem about running permutation or randomisation tests for even very large data sets, and such methods are in very common use today. This is yet another example of Fisher's far-sighted genius!

Observational data versus experimental data

At this juncture, we need to briefly consider how data typically arise in practice. Many research studies are conducted on what is generally termed *observational* data, namely data that have been collected 'in the field' on individuals that are available for the study in question. Of course, statistical principles such as those underpinning valid sampling from target populations and protection against violation of distributional assumptions may well be scrupulously adhered to, but the investigator has no control over many factors when collecting such data and has to be content with whatever values are measured or observed. This scenario is particularly prevalent in the social sciences, where correlation and regression are consequently very popular methods of analysis. However, while application of these techniques is often a routine matter, care must be taken not to read too much into the results. In many such studies, an ultimate goal is to establish that changes in the explanatory variables *cause* changes in the response measurements, and there is sometimes a misconception that a large correlation between two measurements implies that one of them is in some sense responsible for the other. So we need to highlight some reasons why this belief can be badly misplaced.

The most common way in which a high association between two measurements can be misleading is when they are both associated with one or more other, unobserved, measurements. Sometimes there is a fairly obvious connection with such extraneous variables, as, for example, in a study of school children that showed a high positive correlation between their IQ scores and their weights. The obvious 'hidden' measurement here is the age of each child, since as age increases then so does both the weight (the children are growing) and the IQ score (they are learning, and also perhaps becoming familiar with such tests). No one would therefore seriously entertain this association in any formal causal model, but there are many situations in which the 'hidden' measurement is much less obvious and the danger of false inferences is considerable.

A second potential pitfall arises when two or more variables are *confounded* in their effects on the response measurement of interest. This occurs when these variables have an effect on the response, but there is no way of distinguishing their separate influences. For example, suppose that camshafts are produced by two mechanics, each of whom always uses the same machine, and we discover that the camshafts produced by one mechanic have a significantly larger mean length than those produced by the other. However, it is not possible to pinpoint the mechanics as the cause of the differences, because they may both operate in exactly the same way but the settings of the two machines may be significantly different. Alternatively, the differences may be due partly to differences in mode of operation of the mechanics and partly to differences in settings of the machines. The two factors 'mechanics' and 'machines' are confounded, and we cannot disentangle their separate effects on the camshaft lengths.

So what can we do to improve the possibility of making causal inferences? The best way is to conduct a carefully designed *experiment* in which all the possible factors that might affect the outcome are controlled, confounding is eliminated and extra conditions may be imposed in order to *induce* changes in the response measurements of interest. For example, to avoid the confounding of mechanics and machines, we would need to require each mechanic to produce a set number of camshafts on each machine, and that way we could measure the contributions of both factors to the response (length of camshaft). Of course, it may not always (or even often) be possible to set up a suitable experiment, for various reasons. Some subject areas, for example, agriculture, biology, psychology or medicine are more amenable to experimentation, whereas in those like sociology or criminology, for example, the opportunities are fewer and recourse has perforce to be made to observational data. Also, ethical considerations are increasingly of concern in many situations so may provide obstacles to experimentation. Nevertheless, if it is at all possible then experimentation is to be encouraged, and a considerable body of work now exists on how to best arrange and conduct experiments. We here just give a brief overview of the important points, followed by an outline of the necessary methods of analysis.

Designed experiments

Having developed the analysis of variance as a tool for analysing much of the data that he found at Rothamsted, Fisher then turned his attention to the question of designing experiments in such a way as to make them both efficient and effective in generating the sort of future data that would provide clear conclusions regarding their objectives. Given that he was working at an agricultural station, most of the experiments were concerned with comparing either different *treatments* (such as fertilisers) or *varieties* (say of wheat) when applied to, or grown on, different *plots* of land—so all his work was couched in these sorts of terms. However, it soon became clear that his methods were applicable to many different types of scientific or social investigations (e.g. comparing the effects of different industrial processes on samples of metal, of different treatments for ailments suffered by patients in hospital, of different teaching methods on pupils in schools, and so on), but nevertheless much of the original agricultural terminology has persisted and can still be found in modern text books. So for 'treatment' the reader can substitute any externally controlled condition, while subjects, patients, individuals or units are all synonyms for 'plots'.

The two fundamental principles that Fisher emphasised right from the outset were *randomisation* and *replication*, both being prompted by the large amount of variability that he observed in experimental results. The principle of randomisation, which we have already encountered in connection with samples and surveys in Chapter 2, shares much of the motivation already discussed there. Treatments should always be allocated to plots using some objective random device, not only to avoid any unexpected systematic biases that might be present, but also to ensure that the results from different plots can be considered to be independent of each other. Thus randomisation validates the independent observations assumption underlying the analysis of results and the subsequent inferences. Replication, on the other hand, is concerned with precision: given the amount of variation typically encountered in the field, the assumed common variance σ^2 of the normal populations is usually expected to be high. So if it is to be well estimated, and if any subsequent confidence intervals are to be narrow, we will need large numbers in the samples from each population. Consequently, experiments should be designed so that plenty of repeats of each treatment are catered for.

But of course Fisher was also aware that in agricultural experiments, there are plenty of systematic differences between plots that can be anticipated which need to be allowed for if the experimental results are to be effective, and he initiated the study of experimental design that was to play an important role over the next half century. The most common systematic effect in agricultural experimentation is a fertility gradient in the field, whereby plants will grow better irrespective of treatment or variety when passing across the field in one direction—say from east to west. Thus if all the plants at the west side are given one treatment while all those at the east side are given another, then even if there is no difference in the effects of the treatments the one applied at the west side will appear better than the one applied at the east simply because of this fertility gradient. Such a systematic effect must therefore be removed before treatments can be compared properly. The way to do this is to divide the field up into strips, or *blocks*, that run from north to south, and then to apply all the treatments to the same number of plots in each block. This ensures that all treatments are equally exposed to all fertility conditions, and subsequent comparisons between treatments will not be muddled up with fertility considerations. Such an arrangement is known as a *randomised block design*, and the appropriate analysis of variance adds a line 'between blocks' to the lines 'between treatments' and 'Error'. Examples of blocks abound in all types of experiments: animals from the same litters in animal feeding experiments; patients of similar ages in clinical trials; respondents from the same socio/economic groups in social investigations, and so on.

In fact, the tyre data has such an arrangement, as the three columns of Table 6.1 relate to three different types of car used in the experiment. As there may be systematic differences between car types they should be treated as blocks, so there are now three replicates for each car position for each block. The effect of removing this systematic source of variation is to obtain a sharper estimate of σ^2, as the between-blocks sum of squares comes out of the error sum of squares since the comparisons between the car positions are unaffected, and enables differences between car types to be tested as well as differences between tyre positions. To give the details briefly, the previous 'within-groups' sum of squares 923.12 on 32 degrees of freedom breaks down additively into a 'between car types' value of 156.88 on 2 degrees of freedom and an 'Error' value of 766.24 on 30 degrees of freedom. Thus the error mean square shows that the estimate of σ^2 is reduced from 28.85 to 25.54. This enhances the F ratio for 'between tyre positions' (previously given as 'between groups'), and shows from the F ratio for 'between car types' that there is some difference between the three car types.

Once the general idea was established, various other designs were developed for removing other types of systematic variation. For example, if there are fertility gradients in both east–west and north–south directions, then a *Latin square* design will be appropriate, in which there is an equal representation of the treatments in strips running parallel to *both* directions; if blocks are too small to allow all treatments to be represented in each block then the *balanced incomplete block* design has every *pair* of treatments occurring the same number of times across all blocks; if the number of treatments is very large (as in variety trials, for example) then *lattice* designs have similar conditions on pairs of treatments occurring together; and if a further set of treatments needs to be introduced after an experiment has already been started then the *split plot* design can be used. For all of these designs there are fully developed analysis of variance templates, and books on experimental design give all the necessary plans and details of analysis.

The above design considerations apply to the structure of the plots (or samples, patients, subjects, etc.), but frequently there is also a structure in the treatments. The prototype in the agricultural context is in fertiliser trials, where the treatments consist of mixtures of different amounts of nitrogen (N), phosphorus (P) and potassium (K) respectively. The technical terms used here are *factors* for the chemicals and *levels* for the amounts of each chemical in the mixture. The old-fashioned approach to experimentation was to vary the levels of one factor keeping the other factors constant, in order to estimate the effect of each factor level, but this did not give any indication as to how the factors might *interact* with each other. The statistical revolution in this area was to develop the concept of *factorial designs*, in which the full set of possible combinations of levels of each factor are represented in a single experiment. For example, if we wish to study the effects of three different levels (amounts) of each of N, P and K then a single replicate of the experiment would need $3 \times 3 \times 3 = 27$ plots to accommodate all combinations of levels of each factor. Now it is the treatment sum of squares that has to be partitioned rather than the error sum of squares, and it is partitioned into the *main effects* of each of the factors (the effects of the different levels of each factor averaged over all the levels of the other factors) and the *interactions* between pairs, triples, and so on of the factors (the extent to which the joint action deviates from the sum of the separate actions). It may also happen that a full combination of all levels of all factors would require too many plots, so reduced designs known as *fractional replications* can be used if some information on interactions is sacrificed. Factorial designs, and particularly fractional replications, have become very important in industrial experimentation in recent years. Again, all relevant details of the plans and analyses are given in textbooks on experimental design.

The general linear model

It might seem from the foregoing that analysis of variance as obtained in regression studies, where we assume that the y values are made up of a mean given by the linear model plus an error drawn from a normal population, and analysis of variance in designed experiments are two rather distinct techniques. Indeed the early development accentuated the distinctions, because the analysis of variance in experiments was developed very much on a design-by-design basis. However, in fact there is just one single underlying principle, and the analysis of variance for *any* experimental set-up can be recovered from that of regression analysis by appropriate formulation of a linear model. To show the general idea, consider several scenarios built round the tyre data.

First, suppose that only Front (F) or Back (B) has been recorded as the tyre position, and no distinction has been made between Nearside and Offside. Thus the 18 observations in the top half of Table 6.1 are from population F, while the 18 in the lower half are from population B. The means of these two sets of 18 values are 19.57 and 30.14 respectively. Is this evidence that back tyres in general show different amounts of wear from front tyres? To answer this question we need to test the null hypothesis H_0: $\mu_1 = \mu_2$ against the alternative H_1: $\mu_1 \neq \mu_2$, where μ_1 and μ_2 are the population mean reductions in tread depth for front and back tyres respectively. General methodology for such tests has been discussed in Chapter 4 and, as we have already commented earlier, the specific technique in this case would be a two-sample t-test.

However, suppose we define a new variable x for all tyres, and allocate the value 0 on this variable to every front tyre and the value 1 to every back tyre. Let us denote the reduction in tread depth (irrespective of tyre position) by y. Then all 36 tyres now have a *pair* of values: y_i for the ith tyre's reduction in tread depth, and x_i which indicates whether that tyre was at the front or back of the car.

Since we have two variables x and y, we can conduct a simple regression analysis of y on x and obtain the fitted regression equation $y = \hat{a} + \hat{b}x$. Thus the regression analysis will estimate the mean reduction in tread depth $a + bx_i$ for a tyre whose x value is x_i. But all front tyres have value $x_i = 0$, so all their means are a, while all back tyres have value $x_i = 1$ so all their means are $a + b$. Thus estimating the regression coefficients a and b is equivalent to estimating the overall mean tread depths, while testing the worth of regression, namely testing the hypothesis $b = 0$ against the alternative $b \neq 0$, is equivalent to testing whether or not there is a difference between the front and back population means. So the ANOVA test of worth of regression is exactly equivalent to the two-sample t-test.

Suppose next that, instead of looking at differences between tyre positions, we look at differences between car types. The three columns in

Table 6.1 relate to three different car types, so let us call them A, B and C. To investigate differences between these three types, we can use the between-group/within-group ANOVA discussed above. But equally, we can now define *two* extra variables x_1 and x_2 (usually either called *indicator* or *dummy* variables) and assign values to them as follows: $x_1 = 0, x_2 = 0$ for all cars of type A, $x_1 = 1, x_2 = 0$ for all cars of type B and $x_1 = 0, x_2 = 1$ for all cars of type C. Then we fit the multiple regression model $y = \hat{a} + \hat{b}x_1 + \hat{c}x_2$ to the data, and use the resulting ANOVA to assess the significance of the regression.

Consider the y means implied by this model, given the values of x_1 and x_2 for each car. These means will be a for all cars of type A, $a + b$ for all cars of type B and $a + c$ for all cars of type C. Thus the ANOVA test of worth of regression is equivalently a test of differences among the three group means. If the regression is *not* significant then there is no evidence against the hypothesis $b = c = 0$, so there is no significant difference among the three car types. If the regression *is* significant then at least one car type differs from the others.

This reasoning can be extended to any number of groups. If there are k such groups then we simply need to define $k - 1$ indicator variables appropriately, and conduct a multiple regression analysis. In this case the 'regression sum of squares' RSS can be shown to be formally the same as the previous 'between-group sum of squares' BSS, and the analysis of variance is sometimes referred to as a 'one-way analysis of variance'. Thus if we look at the original four tyre positions (FO, FN, RO, RN), we would need to define the three indicator variables x_1, x_2, x_3. This can be done in a variety of ways, but following the three car-types example we could set $x_1 = 0, x_2 = 0, x_3 = 0$ for FO tyres; $x_1 = 1, x_2 = 0, x_3 = 0$ for FN tyres; $x_1 = 0, x_2 = 1, x_3 = 0$ for RO tyres; and $x_1 = 0, x_2 = 0, x_3 = 1$ for RN tyres. Multiple regression analysis will then generate the values in the ANOVA of Table 6.2. Of course, the entries in this ANOVA table can be written down directly and in much simpler fashion using the between- versus within-group formulation than by using the multiple regression formulation, so the simpler version is preferred in textbooks.

By extension, the multiple regression approach with indicator variables can be applied to all the experimental designs mentioned above. For example, in the randomised block design we would need to introduce one set of indicator variables for the blocks and a separate set of indicator variables for the treatments. Thus to incorporate car types *and* tyre positions for the data of Table 6.1, we would need the three dummy variables x_1, x_2, x_3 defined in the previous paragraph for the four tyre positions, and then a further two dummy variables x_4, x_5 whose values are defined as in several paragraphs earlier for the three car types A, B and C. In general with multiple regression, it has already been stressed that if we want to test the significance of a particular set of variables then we need to find the 'extra sum of squares' due to this set *after* fitting all the other variables first. Thus, if the objective is to test for differences in tyre positions, then the sum of squares for the car types would first be obtained from the regression sum of squares on the model that includes the variables x_4, x_5, and the sum of squares for the tyre positions would then be the 'extra sum of squares' on adding the variables x_1, x_2, x_3 to this model. But if the objective is to test for differences in car types, then we would first need to fit the model formed from the variables x_1, x_2, x_3 and then obtain the sum of squares for the car types as the 'extra sum of squares' on adding x_4, x_5.

However, the great benefit of all the classical experimental designs mentioned above is their *balance*: equal replication of each treatment in each block, equal occurrence of all pairs of treatments in balanced incomplete blocks, equal number of plots carrying each level of each factor in factorial designs, and so on. This balance ensures that there is orthogonality between the sets of indicator variables corresponding to each effect of interest, and this in turn is what makes the ANOVAs for these designs so simple to formulate. Moreover, once we move to the regression formulation, this orthogonality means that the order in which we fit the variables is immaterial; we always get the same numerical results. So, for example, with the tyre data, the sum of squares due to the regression on just x_1, x_2, x_3 is exactly the same as the 'extra sum of squares' of x_1, x_2, x_3 after fitting the model that already includes x_4, x_5. Thus all the terms in the ANOVA can be easily obtained by fitting just a few (relatively simple) regressions.

So the reader might finally wonder why we have considered the regression formulation at all—would it not just be far simpler to keep to the ANOVA formulae for the appropriate design? Well yes it would—provided that our data conform exactly to such a design. Real life is never that accommodating, however, and in many studies things go wrong: crops fail on some plots in agricultural experiments; animals die during feeding trials; patients in a clinical trial give up on the treatment for various reasons; the researcher conducting the experiment loses some of the results, and so on. In spite of the best intentions, and notwithstanding a very carefully designed experiment, the end product can be far from what was envisaged at the outset. Once the design balance is lost for any reason, such as the ones above, then the effects lose their orthogonality, the simple ANOVA formulation is no longer appropriate and recourse must be taken to the regression approach (with all the attendant complications of non-orthogonal models). Moreover, the regression approach is always available irrespective of how the data have been collected, so will cope with something that seems to be a sensible design arrangement but isn't catered for in standard textbooks. Most computer packages contain a set of subroutines or modules that enable these analyses to be carried out. These subroutines are commonly found under the heading 'the General Linear Model', making it perhaps one of the most useful sections of the package.

7 Generalising the Linear Model

Introduction

We start by reviewing the most general form of the linear model, namely the multiple regression model, but stating it in a slightly more formal way than we did in Chapter 6. This model is the simplest one for studying the dependence of a response variable y on a set of explanatory variables $x_1, x_2, \ldots, x_k$. For a full analysis (including not only parameter estimation but also tests of hypotheses) we need to make a very definite assumption, namely that if we observe a particular value y_i of the response variable, and the associated values of the explanatory variables are $x_{i1}, x_{i2}, \ldots, x_{ik}$, then y_i is in fact a value from a normal population whose mean is $a + bx_{i1} + cx_{i2} + \cdots + dx_{ik}$ for some (unknown) parameter values $a, b, c, \ldots, d$. We said nothing in Chapter 6 about the variance of this population, but a further assumption of the technique is that this variance is some constant value σ^2, *irrespective of the value of the population mean* $a + bx_{i1} + cx_{i2} + \cdots + dx_{ik}$. This extra assumption of constant variance is known as the assumption of *homoscedasticity*. An alternative way of phrasing the model is to say that y_i is equal to the population mean $a + bx_{i1} + cx_{i2} + \cdots + dx_{ik}$, plus a 'departure' term that accounts for the random variation and is randomly produced from a normal distribution that has zero mean and variance σ^2. Thus, mathematically, we can write

$$y_i = a + bx_{i1} + cx_{i2} + \cdots + dx_{ik} + \varepsilon_i$$

where ε_i represents this departure term. But there is one further assumption, which is implicit in all the methods that we have come across so far, and that is that if we now observe a sample of n individuals then their resulting values $y_1, y_2, \ldots, y_n$ are all mutually independent. In the formulation above, this translates into the equivalent assumption that the corresponding ε_i are independent of each other. So, in summary, we have the following four essential assumptions in the multiple linear regression model:

(1) the mean μ_i of y_i is a function of the explanatory variables that is linear in its parameters;
(2) the variance of y_i is a constant value σ^2 for all i;

(3) the distribution of y_i is normal; and
(4) the individual y_i are mutually independent.

In any practical application, one or more of these assumptions may not be appropriate. Sometimes it is evident right from the start that this is the case. For example, when studying the rate of chemical reactions by collecting data on the rates under different settings of temperature and pressure, it may be known from chemical theory that a non-linear model linking these quantities is to be expected. Or when collecting data on incidences of an infectious disease across a county in the United Kingdom, it is reasonable to suppose that incidences in nearby towns will be more similar than incidences in a pair of distant localities, so the assumption of independent observations is questionable. We would therefore be looking for more appropriate models to fit to the data right from the outset of these studies. In other situations we may not have any prior theoretical or practical knowledge to guide us and we are simply conducting an exploratory modelling exercise. In this case we would like to find the simplest model that will explain the phenomenon under study, and the multiple linear regression model is the obvious place to start. So we need to have some diagnostic procedures with which we can first check whether the assumptions of the technique are being satisfied, and which will ring an alarm bell if they are not. While this warning does not necessarily mean that the analysis is invalidated, nevertheless it does mean that a better technique or analysis can usually be found.

The purpose of this chapter is therefore to briefly review the thinking behind various diagnostics for checking assumptions, and then to look at generalisations of the multiple linear regression model that will allow for deviation from each of the four assumptions above. Unfortunately, many of these generalisations take us either into the realms of complicated mathematics or into equally complicated computational algorithms. Indeed, although some of the possible generalisations have been known about for quite a long time, it was not really until the last quarter of the twentieth century, when computing power increased dramatically, that it became feasible to conduct them on a routine day-to-day basis. So we will be able to go into even less technical detail than we have managed so far, but we hope that by focussing on the underlying reasoning and logic the reader will develop at least an understanding of the aims and procedures of each generalisation. All the techniques are now readily available on standard statistical software packages, such as Minitab, R, Splus, SPSS, SAS, GENSTAT and others, so there is no problem about conducting the analyses.

Checking the linear model assumptions

Suppose that we have a set of data as above, and we fit the multiple regression model

$$y_i = a + bx_{i1} + cx_{i2} + \cdots + dx_{ik} + \varepsilon_i$$

after making all the necessary assumptions. The method of least squares provides us with estimates $\hat{a}, \hat{b}, \hat{c}, \ldots, \hat{d}$ of the parameters, and the fitted values $\hat{y}_i$ are given by

$$\hat{y}_i = \hat{a} + \hat{b}x_{i1} + \hat{c}x_{i2} + \cdots + \hat{d}x_{ik}.$$

The *residuals* are defined to be the differences $y_i - \hat{y}_i$ between the observed y_i and their corresponding fitted values, so it is easy to see that they are in fact estimates of the model departures ε_i. Therefore, roughly speaking, *if the correct model has been fitted and all model assumptions have been obeyed*, the behaviour of the residuals should be similar to the ideal behaviour of the ε_i. Strict theory shows that this will not be quite true, but for all practical purposes it is approximately the case. Although many computational procedures have been devised in order to test formally for departures from the various assumptions, appropriate plots will often show up potential departures in a very simple fashion so it is on such plots that we focus here. If there has been no violation of the four assumptions then the values of the residuals should be consistent with those of a sample from a normal population with constant variance. Probability plotting will check out the normality assumption, and a plot of the residuals against any other variable should show a random scatter about the value zero, of approximately constant 'width' across the range of the other variable if the other model assumptions are satisfied. Plots that show systematic patterns deviating from this behaviour indicate violation of some or all of the assumptions, so we now consider briefly each assumption and how to detect its violation.

Non-linearity of model

Figure 7.1 shows a plot of 15 values of the non-linear function $y = e^{0.5x}$, with x taking values $1, 2, 3, \ldots, 14, 15$, and superimposed on the plot is

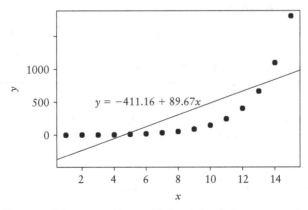

Figure 7.1. Exponential response data, with fitted simple linear regression.

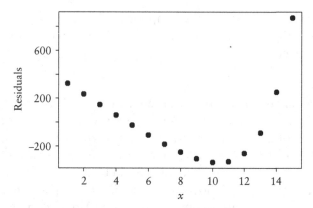

Figure 7.2.   Plot of residuals against x for data from Figure 7.1.

the straight line $y = -411.164 + 89.667x$ fitted by simple linear regression to these points.

For x from 1 to 4, the line is below each point, so the fitted $\hat{y}_i$ is less than the actual value y_i in each case and therefore the four residuals $y_i - \hat{y}_i$ are positive. Then for x from 5 to 13 the line is above each point, so the residuals are all negative. Finally, for x at 14 and 15, the points are again above the line so the residuals revert to being positive. Thus if we plot the residuals against the x_i, as in Figure 7.2, then we clearly see this systematic non-linear pattern. Such a pattern would indicate that some non-linearity was missing from our model and that the simple linear regression fit was not appropriate. Of course, it may not be immediately obvious what the most appropriate amendment should be, and we might need to try some alternatives (e.g. add a quadratic term in x, fit an exponential model, etc.) before hitting upon the correct model. However, the alarm bell has been sounded!

Non-constant variance

Constant variance implies that when the residuals are plotted against any other variable, the band of points across the plot should be of roughly constant width. Any systematic departure is indicative of heteroscedasticity, but it is often difficult to judge whether there is systematic departure or not. The most common cause of such departure is dependence of the variance on the value of one of the explanatory variables, or on the mean of the response variable (i.e. on the systematic part of the model), and plotting residuals against either an explanatory variable or the fitted values should show up any dependence.

As an illustration, 40 observations were generated from the model

$$y_i = 2.0 + 0.5x_i + \varepsilon_i,$$

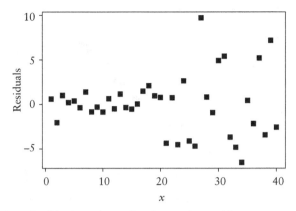

Figure 7.3.   Plot of residuals against x for observations with non-constant variance.

at values 1, 2, 3, . . . , 39, 40 of x. For the first 20 observations, the departure terms ε_i were randomly generated observations from a normal distribution having mean zero and variance 1.0; while for the remaining 20 observations, they came from a normal distribution having mean zero and variance 16.0. A simple regression was fitted to the data, the fitted equation was $\hat{y} = 1.795 + 0.472x$, and Figure 7.3 shows a plot of the residuals against the values of x. It is evident that the residuals for values of x greater than 20 are much more widely scattered about zero, that is, have greater variance, than those for values of x less than 20.

These data had essentially two groups of observations, the first with smaller variance than the second, so that the plot showed a dichotomy at somewhere around $x = 20$. However, this dichotomy was so marked because the variances in the two groups were very different, and in many practical situations it may not be so easy to detect such a shift. If instead of a dichotomy the variance had depended directly on the value of x, then the plot would appear more 'funnel-like', with the scatter of points being tight at small values of x (so small variance) and then gradually fanning out to large scatter at large values of x (and hence large variance). The dichotomy and the 'funnel' are the two most common types of heterogeneity of variance in practice, but note that in many cases it may be difficult to distinguish between them and all that one might be able to say is simply that the variance is not constant across all observations.

Non-normality

The idea of a probability plot is very simple: if the sample values are arranged in increasing order and plotted against the appropriate scale for a particular distribution, then for a random sample from that distribution the plot should approximate to a straight line. The most convenient way of obtaining the appropriate scale is to find the expected values, from the relevant distribution,

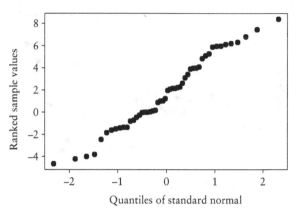

Figure 7.4. Plot of ordered sample values against standard normal quantiles for a sample from a normal distribution.

of an ordered sample of the same size as the observed sample. Note that both of these sets of ordered values are *quantiles*, the first set being the sample quantiles and the second set being the postulated population quantiles. Hence such a plot is known as a *quantile-quantile*, or Q–Q, plot.

If the target distribution is the standard normal, then these expected values are known as the *normal scores*, and they are readily obtained in most statistical packages. So a suitable graphical check on the normality of the set of residuals (or, alternatively, of the response values y_i) is to arrange them in ascending order, generate their corresponding normal scores, and then plot the ordered values against the normal scores. If the sample values do indeed follow a normal distribution then the plot will be approximately linear (to within the usual tolerance of random variation). To illustrate what we mean by 'approximately linear', 50 observations were randomly generated in Splus from a normal distribution with mean 2.0 and variance 9.0, and the Q–Q plot of Figure 7.4 was produced. Splus has a command that produces the plot directly, but other packages may not. To produce the equivalent plot in one of these packages, for example, Minitab, the sample values would first need to be arranged in ascending order, the corresponding normal scores would need to be obtained using the Minitab command NSCORES, and the ordered sample values would need to be plotted against their normal scores. Although the points wobble about a bit, it is clear that if a straight line were fitted to them then they would all lie very close to it.

However, if the sample values are not from a normal distribution then the plot will show some systematic deviation from linearity. The nature of this deviation will reflect the nature of the non-normality. One possibility is that the distribution is very different from a normal distribution, for example, it is highly asymmetric, in which case there will be some pronounced non-linearity to the plot. An illustration of such a deviation is shown in Figure 7.5, where an ordered sample of 50 observations that were randomly generated

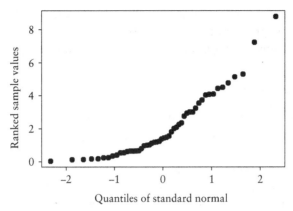

Figure 7.5. Plot of ordered sample values against standard normal quantiles for a sample from an exponential distribution.

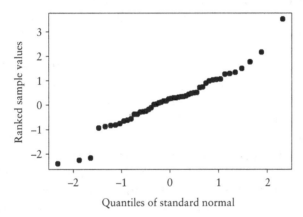

Figure 7.6. Plot of ordered sample values against standard normal quantiles for a sample from a t-distribution on 4 degrees of freedom.

from an exponential distribution with mean 2.0 is plotted against the standard normal quantiles. Although the mean of the sample is the same as before, the exponential distribution is highly asymmetrical and this is shown up by the pronounced curvature of the plot.

A common situation is when the distribution is approximately symmetrical but either 'flatter' (i.e. with heavier 'tails') or more 'peaked' than the normal. Figure 7.6 illustrates this by showing a plot of the standard normal quantiles against the ordered values from a t-distribution on 4 degrees of freedom. This distribution has much heavier tails than the normal, which means that it will tend to have more outlying values, so the plot will typically be approximately linear in its middle portion (reflecting the 'normal-like' symmetry) but with several points at either end that lie away from the line of the remainder (reflecting the outlying values). This pattern is indeed shown in Figure 7.6, with one outlying value at the upper end and three at the lower end.

There are very many formal statistical tests of normality in the literature, but the normal probability plots illustrated above provide an adequate guide for checking this assumption in most practical investigations.

Non-independence

The most common form of non-independence of observations is when *serial correlation* is present, which is when the successive values presented for regression analysis are correlated. This happens most often when the observations are made successively in time, and the value of the observation at one time point affects the value observed at the next time point. An obvious example is the measurement of temperature at a particular place. If the temperature is measured every hour, then the value will not oscillate randomly from observation to observation. Clearly, there will be some systematic pattern with values gradually rising and then falling throughout the day. Other situations giving rise to serial correlation are when measurements are made at contiguous points in space, or on individuals that have some natural connection.

In order to detect this systematic trend numerically, suppose that the variable y denotes the set of values for analysis, and another variable z denotes the same set when shifted back by one (so that if the y values are, for example, 6.4, 7.2, 7.8, 8.4, 7.5, 7.0, 6.7 then the z values are 7.2, 7.8, 8.4, 7.5, 7.0, 6.7). If we omit the last y value 6.7 (so that y and z have the same number of values) and find the correlation between the resulting (y, z) pairs of values [(6.4, 7.2), (7.2, 7.8), and so on], then we are finding the serial correlation between each observation and its immediate successor. A large positive serial correlation implies high dependence between successive values, a large negative serial correlation implies an oscillatory set of values and a serial correlation close to zero indicates no dependence between successive values.

To check for this type of non-independence among the residuals we would either need to plot them in the order in which they were collected and observe that there was a more systematic trend in their values as opposed to random fluctuation (typically a slowly varying pattern in which positive residuals are grouped with other positive ones and negative residuals with other negative ones), or to test their serial correlation for significance, or to use one of a number of nonparametric tests (such as the 'runs' test) for serial correlation that are provided in most standard statistical packages. These tests assume that the order in which the observations are entered into the computer is the order in which they were collected.

So the above has provided a brief summary of the diagnostic checks we can apply to investigate appropriateness of each assumption in the fitting of linear models. We now look at what we can do in case each assumption fails.

Non-constant variance: weighted regression

Suppose that we either know that the variances of our response variable values y_i are not all equal, or one of our diagnostic plots has suggested that this is the case. Then we can say in general that the variance of y_i is σ_i^2, which depends on the actual observation i, instead of the constant σ^2. What can we do to remedy the situation? Providing that all our other necessary regression assumptions are satisfied, the solution in theory is very straightforward. When the variance is constant, then all observations are subject to the same 'error' in their measurement, and the least squares principle reflects this fact by giving each observation an equal part in the estimation process. However, when the variance is different from observation to observation, then those observations with high variance are subject to larger 'error' in their measurement than the ones with low variance. If the variance reflects 'error', then the inverse of the variance reflects 'precision' and, clearly, the more precise observations (the ones with smaller variance) should be given more influence in the estimation process. So, instead of finding the parameter estimates $\hat{a}, \hat{b}, \hat{c}, \ldots, \hat{d}$ by minimising the straightforward sum of squared differences between the y_i and their predicted values $\hat{y}_i = \hat{a} + \hat{b}x_{i1} + \hat{c}x_{i2} + \cdots + \hat{d}x_{ik}$, a better procedure is to find the estimates that minimise the sum of *weighted* squared differences between the y_i and their predicted $\hat{y}_i$, where the weight w_i attached to observation i is the inverse of its variance, that is, $w_i = 1/\sigma_i^2$.

In practice, of course, we will never know these variances σ_i^2 exactly, but that is not a drawback because we only have to know them up to a constant of proportionality. In other words, provided that we can say that $\sigma_i^2 = \sigma^2 k_i$, where we know k_i but not σ^2, then we use $w_i = 1/k_i$. So how can we obtain information about k_i? Sometimes it happens that we can deduce the values of k_i from theoretical considerations. For example, if the observation y_i is actually the mean of n_i values, each having variance σ^2, then from Chapter 4 we know that the variance of y_i is σ^2/n_i (the larger the number of values contributing to the mean, the smaller its variance and hence the greater its precision). In this case, $k_i = 1/n_i$ and hence $w_i = n_i$, which clearly makes sense as the greatest weights should be accorded to those y_i values that have been computed from the most observations.

To illustrate the worth of such an approach, observations were produced from the model

$$y_i = 2.0 + 0.5x_i + \varepsilon_i$$

at values $1, 2, 3, \ldots, 19, 20$ of x, by randomly generating values ε_i from a normal distribution having mean zero and variance 1.0. Moreover, n repeat observations were independently generated at the value $x = n$ (i.e. 1 observation at $x = 1$, 2 observations at $x = 2, \ldots, 20$ observations at $x = 20$)

and the mean of these replicates provided the y value at $x = n$. Thus according to the above argument, a weighted regression that gives weight n to the observation at $x = n$ should be preferable to an ordinary regression for these data, as the variance of the observation at $x = n$ is $1/n$. In fact, the fitted regressions using Splus were $y = 3.065 + 0.420x$ for ordinary regression and $y = 2.624 + 0.452x$ for the weighted regression. Thus both estimated regression parameters are closer to their true values with the weighted regression, and this approach is indeed the better one.

In general, however, we may not have such theoretical information so may have to rely on interpretation of the various diagnostic plots above for deciding on the w_i. For example, if a 'funnel-shaped' plot suggests that the variance increases with the values of a particular explanatory variable x, then the inverses of the x values will provide a suitable set of weights. If the plot suggests that one group of observations has a variance approximately twice the size of the rest of the observations, then the weights of observations in this group should be one half of the weights of the rest, and so on.

Once a set of weights has been decided, the actual conduct of the regression analysis is very straightforward. Mathematical theory establishes that all we need to do is to multiply all the variable values (including the dummy variable having a constant value 1 corresponding to the regression intercept) for a particular individual by the square root of that individual's weight, and then to conduct standard regression analysis on the resulting variables. So we can obtain the full gamut of results without difficulty, and indeed all standard regression software packages include an option for specifying a set of weights and carrying out the weighted regression.

A particularly interesting case of weighted regression occurs when the variance of y_i depends on its mean, namely on the value of $a + bx_{i1} + cx_{i2} + \cdots + dx_{ik}$. Since an estimate of this mean is given by $\hat{y}_i = \hat{a} + \hat{b}x_{i1} + \hat{c}x_{i2} + \cdots + \hat{d}x_{ik}$, a funnel shape is to be expected when residuals are plotted against fitted values $\hat{y}_i$. If the funnel is such that the variance rises as $\hat{y}_i$ increases then the above discussion would suggest using weights $w_i = 1/\hat{y}_i$; while if the reverse is the case, and the variance falls as $\hat{y}_i$ increases, then the weights should be $w_i = \hat{y}_i$. So we would carry out a weighted regression with the appropriate weights. However, the matter would not generally rest there, because the new regression yields *new* fitted values $\hat{y}_i$ and hence, in general, a new set of weights. This leads to a yet further weighted regression, and a repeat of this cycle. This process has been termed *iteratively reweighted least squares*, and it occurs behind the scenes in some of the generalisations below as well. Usually only a few such cycles are required before the system settles down ('converges' in computational parlance) and gives the same fitted values and hence weights at the end of each cycle. This is the point at which the process can be stopped and the results can be reported.

Non-linear models

The multiple regression model considered so far, namely

$$y_i = a + bx_{i1} + cx_{i2} + \cdots + dx_{ik} + \varepsilon_i,$$

implies that the response variable y is linearly related to each of the explanatory variables $x_1, x_2, \ldots, x_k$. However, in many practical situations a simple plot will readily reveal some non-linearity in the relationship between y and one or more of the xs, so for accurate analysis we must consider different models. An observed curvilinear relationship between two variables may perhaps be adequately modelled by a quadratic, $y = a + bx + cx^2$ say, and we have already demonstrated that such a relationship can be fitted by ordinary least squares since it is linear in the unknown parameters. The same is true of a more complicated polynomial, such as the pth order one $y = a + bx + cx^2 + \cdots + dx^p$, as this is also linear in its parameters so can be fitted by multiple regression on defining the variables $x_1 = x$, $x_2 = x^2$, $x_3 = x^3, \ldots, x_p = x^p$. It might seem, therefore, that the easily implemented multiple (linear) regression can also be used to fit most curvilinear relationships, but this is illusory. It is often the case that high-order polynomials would be needed in order to provide an adequate fit to observed data, but fitting them is not usually a good idea; such models are not only very volatile to fit (because of the large number of possibly highly correlated powers of x), but prediction from them can also be disastrous (because of their induced oscillatory behaviour). Moreover, some relationships simply cannot be represented by a polynomial (e.g. ones that are not symmetrical about their optima, or asymptotic ones that gradually approach a 'floor' or 'ceiling'). So in general, if it seems that a polynomial of higher than third order is needed, it is preferable to move to non-linear regression.

Following our previous definition, a model that is non-linear in its parameters means that a unit increase in one or more parameters produces changes in the response variable that differ according to the values of the explanatory variables. For example, consider the allometric (growth) model $y = ax^b$, where the parameters are a and b. If a is changed to $a + 1$, the corresponding change in y is $(a + 1)x^b - ax^b = x^b$, which is different at every different value of x. Various other non-linear forms are commonly used in practice, including the exponential ($y = ae^{bx}$), the inverse polynomial ($y = 1/(a + bx)$), and the logistic ($y = a/(1 + e^{b+cx})$). Some specific examples are shown below: Figure 7.7 shows the exponential function $y = 3.2e^{0.25x}$, Figure 7.8 shows the inverse polynomial $y = 1/(2 + 0.5x)$, while Figure 7.9 shows the logistic function $y = 1 - (1/(1 + e^x))$. The exponential typically has a steeper rise or decline than the inverse

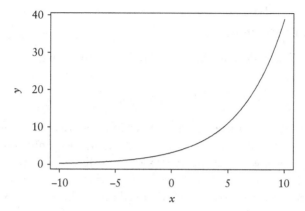

Figure 7.7.   Plot of an exponential function.

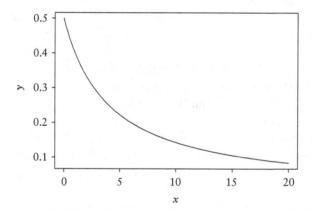

Figure 7.8.   Plot of an inverse polynomial.

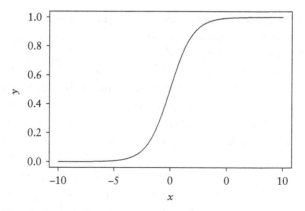

Figure 7.9.   Plot of a logistic function.

polynomial, while the logistic exhibits the *sigmoid* shape (suggestive of an elongated 'S') that is characteristic of cumulative probability distributions and hence useful for modelling probabilities.

These forms would typically be used to model the means of the observed response variable values y_i, with an additive departure ε_i representing the random variation. For example, an exponential model with a single explanatory variable x would be written

$$y_i = ae^{bx_i} + \varepsilon_i,$$

where the ε_i are subject to the usual assumptions as set out earlier. Extending such models to include more explanatory variables is straightforward, for example,

$$y_i = ae^{bx_{i1} + cx_{i2}} + \varepsilon_i,$$

if there are two explanatory variables.

One of these models will often represent the physical process under study much better than a polynomial model, but they are all more complicated to fit. Essentially, we still use least squares to estimate the parameters; in other words, we find the values of the parameters by minimising the sum of squared residuals $y_i - \hat{y}_i$. But whereas for linear models these parameter values can be found mathematically in terms of the data, and only need numerical values to be substituted by the computer (a very simple and fast operation), this is no longer the case for non-linear models. Parameter values have to be found numerically, with the computer effectively having to search among all possible values of the parameters and compute the fitted values at each parameter combination in order to find the combination that minimises the residual sum of squares. Of course there are various strategies (numerical methods and computer algorithms) for systematising and speeding up this search, but in some cases it can be a laborious iterative process and for some data sets it can even fail entirely.

Before the advent of fast computing power, therefore, much effort in any study was frequently expended on seeking transformations of the variables that would 'linearise' the model and enable standard linear regression to be used. The exponential model provides a simple example, where taking natural logarithms (ones to base e) converts the non-linear form $y = ae^{bx}$ to the equivalent linear form $\log y = \log a + bx$. Thus it seems that fitting a linear regression of $\log y$ on x will provide a much simpler solution. However, the problem is that such transformations usually focus exclusively on the response variable means, and the departure terms are ignored. Thus, although the systematic part may be suitably linear, the assumptions about the departures are often not satisfied. Here, in fitting the linear version by simple regression we assume that the departures from the logarithms of the y values are additive. But the sum of logarithms of a set of values is the logarithm of their product, so additive departures in $\log y$ implies *multiplicative* departures in y. This in turn implies non-homogeneous variance in y, the variance increasing dramatically as y increases, which may not be at all appropriate for the data in hand. Given the present-day cheapness and speed of computing, it is generally preferable now to fit non-linear models. However, one drawback that should be borne in mind is that any significance levels or confidence intervals that are derived as part of the analysis will generally only be approximate ones.

It can also be noted that since the advent of fast computing power, many nonparametric computer-intensive variants of regression have also been developed for fitting non-linear curves to data. Instead of postulating a fixed form of model, as in the cases above, these methods all let the data 'speak for themselves' in determining the appropriate curve. They are essentially 'local' fitting procedures, in that they either gradually move through the data set, fitting simple functions such as low-order polynomials as they go, or they fit these separate functions to small sections of the data and ensure that they are adequately 'joined up' to provide a single optimal curve for the data. The former approach is known as 'smoothing' the data, while the most common methodology of the latter is spline regression; associated techniques include generalised additive models, projection pursuit regression and regression trees.

However, most of these techniques usually require specialised software and some non-trivial technical knowledge for their successful implementation.

Non-normality: generalised linear models

All the models we are considering have the general form

$$y_i = \mu_i + \varepsilon_i,$$

where y_i is a particular value of the response variable we are interested in, μ_i is the mean of the response variable that is some function (either linear or non-linear) of the values $x_{i1}, x_{i2}, \ldots, x_{ik}$ of a set of explanatory variables, and ε_i is the departure of the particular y_i from this mean. The mean is often called the *systematic* component of the model, while the departure is the *random* component. All the methods of regression analysis discussed so far have assumed the random component to have a normal distribution. This assumption implies that the y_i values come from normal distributions whose means depend on the explanatory variables, but in many practical situations this is clearly not an appropriate assumption and one of the distributions discussed in Chapter 3 would be more appropriate. Some examples are as follows.

In insecticide trials, different batches of insects are sprayed with varying amounts of the insecticide and the proportions of insects surviving in each batch after a given number of days are recorded. If we want to investigate the dependence of proportion of insects surviving on dose of insecticide, a binomial assumption for the proportions is clearly much more appropriate than a normal assumption.

In a study of wave damage to cargo ships, the numbers of damage incidents were recorded for ships of five different types, built at four different time periods and of given aggregate months' service. These numbers were of course integers, and ranged from 0 to 58. To investigate the dependence of damage incidents on ship type, date of building and aggregate months' service, a Poisson assumption for counts would be much more reasonable than a normal assumption.

Many studies of rainfall have been conducted in different parts of the world as part of agricultural investigations, in order to explore how the amount of rainfall depends on season and local environmental conditions. All these studies support the assumption of a gamma distribution, rather than a normal assumption, for the distribution of rainfall amounts.

In all such studies, the use of standard regression analysis will make the wrong distributional assumption and hence significance tests and confidence intervals (all dependent on normality) might be badly wrong. In addition, when data are non-normal then the variance of the response variable is generally not constant, and fitted values or predictions of the response variable may not make sense (e.g. fitted values that are either less than zero or greater than 1 for a proportion). So alternative methods are necessary.

The *generalised linear model* was developed in the 1970s, starting with a seminal paper by John Nelder and Robert Wedderburn in the *Journal of the Royal Statistical Society Series A* (1972) to provide such extensions to ordinary regression, and it has a form that allows many of the familiar regression interpretations to be applied more generally. The essential differences from the linear regression formulation are that the values y_i of the response variable are permitted to come from any distribution in the *exponential family* (which includes the normal, binomial, Poisson and gamma as well as other distributions), and that the linear predictor $a + bx_{i1} + cx_{i2} + \cdots + dx_{ik}$ is applied to a *function* $g(\mu_i)$ of the mean of y_i rather than to μ_i itself. The actual function choice g is dictated by the distribution of y_i. For the normal distribution this function is the identity, that is, the linear predictor is applied to μ_i itself, and the generalised linear model is just the usual linear multiple regression model. For Poisson data, g is the logarithm function, which means that $\mu_i = \exp(a + bx_{i1} + cx_{i2} + \cdots + dx_{ik})$; for binomial data there are several possibilities of which the most common is the logit function, in which case μ_i is a logistic function of the linear predictor, and for a gamma distribution the common choices are either the reciprocal function or the logarithmic function. The function g is usually termed the *link* function of the model.

Procedures for fitting generalised linear models are available in most statistical software packages. The theoretical basis is maximum likelihood rather than least squares (to allow for the different distributions that can now be present), and the generalisation of residual sum of squares is known as *deviance*, but many of the results are expressed in terms familiar from regression analysis. It is both impractical and technically complex to attempt a brief but general description of all the possible models within this class, so we will give a flavour of just one of the generalisations. This is probably the most common of the models in practice, and it gives rise to a technique known as *logistic regression*.

The situation in which this technique is applicable is when the response variable y can take just one of two possible values, traditionally denoted 0 and 1. Thus an example might be the insecticide trial above, in which y denotes an insect's state as either 'alive' (1) or 'dead' (0) and x is the dose of insecticide to which that insect is exposed. In order to determine the effect of dose on chances of survival, we might suppose that an insect's probability of survival π depends on some simple function, say $a + bx$, of the dose it is subjected to, and then use the data to estimate the parameters a, b of this model. Numerical values of these parameters would categorise the relationship between survival probability and dose, while a rejection of the hypothesis $b = 0$ would establish that there was some connection between dose and survival probability.

A more complex example is provided by a study into factors affecting the risk of heart problems conducted by the Health Promotion Research trust in 1986, in which a number of males reported whether or not they had some form of heart trouble (y), and also provided some information about their social class (x_1), smoking habits (x_2), alcohol consumption (x_3), family history of heart trouble (x_4) and age (x_5). If $y = 1$ indicates the presence and $y = 0$ the absence of heart trouble, then we might postulate that the probability π of a male having heart trouble depended on a linear function $a + b_1x_1 + b_2x_2 + \cdots + b_5x_5$ of the explanatory variables. Once again, estimation of the parameters of this model would categorise the relationship between the probability of heart trouble and the values of the explanatory variables, while testing the hypothesis $b_i = 0$ would establish whether or not x_i had any influence (over and above the other explanatory variables) on the probability of heart trouble.

Very frequently, data of this type are presented in *grouped* form. This is where there is a *set* of individuals that all have the same value(s) of the explanatory variable(s). For example, n_i insects are exposed to insecticide dose x_i for each of m doses $x_1, x_2, \ldots, x_m$; or n_i males in the heart study all have the same values of the five explanatory variables for each of m distinct combinations of these variables. Then if r_i of the n_i individuals in the ith set have value $y = 1$, the proportion $p_i = r_i/n_i$ is a simple estimate of π_i for that set's value(s) of the explanatory variable(s). It is convenient to focus on grouped data when discussing the analysis, because ungrouped data are a special case with $n_i = 1$ for all i. However, ungrouped data do have the awkward feature that the only two possible values of each r_i, and hence of each p_i, are 0 and 1, and this has some drawbacks that will be mentioned later.

For the moment, therefore, we assume that we have values of n_i, r_i (hence p_i) and of the k explanatory variables $x_{i1}, x_{i2}, \ldots, x_{ik}$ in set i (for m such sets). The first step in the analysis is to decide on the form of the relationship that determines π_i from $x_{i1}, x_{i2}, \ldots, x_{ik}$. For consistency with linear regression modelling, we maintain a linear predictor $a + b_1 x_{i1} + b_2 x_{i2} + \cdots + b_k x_{ik}$ of the explanatory variables but now there are two crucial differences from linear regression. Equating π_i directly to this linear predictor will sometimes lead to obviously wrong results, as straightforward regression methods can give predicted or fitted values of the probability π_i outside the range $(0, 1)$. Also, this range $(0, 1)$ is essentially non-linear: it is more important in practical analysis to distinguish between similar π_i values at the extremes of the range, say between 0.95 and 0.99 or between 0.05 and 0.01, than it is to distinguish between similarly close values in the middle of the range, say between 0.49 and 0.54. So we need to set the linear predictor equal to some suitable *function* of π_i, and the most appropriate such function has been found to be the *logistic* function $\log_e(\pi_i/(1 - \pi_i))$. This function is also known as logit(π_i), and the effect of the transformation is to stretch the range from $(0, 1)$ to encompass all possible negative and positive values, and moreover to stretch it in such a way as to emphasise differences at the extreme values of π_i relative to those in the middle of their range. So an appropriate model is therefore given by

$$\log_e\left(\frac{\pi_i}{1 - \pi_i}\right) = a + b_1 x_{i1} + b_2 x_{i2} + \cdots + b_k x_{ik},$$

which can be re-expressed (using some simple algebra) as

$$\pi_i = \frac{e^{a+b_1 x_{i1}+\cdots+b_k x_{ik}}}{1 + e^{a+b_1 x_{i1}+\cdots+b_k x_{ik}}}.$$

The final step is then estimation of the parameters $a, b_1, \ldots, b_k$, and, as with all generalised linear models, this is done using maximum likelihood rather than least squares. The essential assumption here is that each r_i has a binomial distribution with parameters n_i and π_i, and all r_i are mutually independent. Thus the likelihood can be written down very easily in terms of π_i, so that substitution for π_i from above gives the likelihood in terms of the parameters $a, b_1, \ldots, b_k$. This allows the likelihood to be maximised over these parameters, using an iterative numerical approach; appropriate routines that do this are provided in most standard statistical software packages, and they provide parameter estimates and standard errors. Several goodness-of-fit statistics of the model can also be derived. One such statistic is based on *deviance* of the model, which is defined as twice the difference between the logarithm of the likelihood at the maximum likelihood estimates of the model parameters and the logarithm of the likelihood when π_i are replaced by p_i. These two logarithms should be similar in value when the model is appropriate for the data, in which case the deviance is 'small', but will differ substantially when the model is not appropriate, in which case the deviance is 'large'. Similar behaviour is encountered with the other popular goodness-of-fit statistic, namely the Pearson X value of Chapter 4 obtained by comparing the observed r_i in each group with the value predicted by the model. Both these statistics have chi-squared distributions under the hypothesis that the fitted model is appropriate, so this distribution provides the mechanism for deciding whether the computed statistic value is 'small' or 'large'. Typical computer package output includes the statistic values, their degrees of freedom and the significance level of the test that the model is appropriate.

Furthermore, facilities are also available for testing the worth of additional explanatory variables, or conducting stepwise variable selection for model choice, much as in multiple regression. The main difference here is that instead of being based on differences in residual sums of squares, these procedures use differences between deviances as the criterion. However, the essential mechanics and interpretations are the same as in multiple regression. The problem with ungrouped data, alluded to earlier, is that for various technical reasons the tests of goodness-of-fit of the overall model are no longer valid. However, there are no problems with tests based on differences of deviances, so the model building process can still be conducted.

Ordinary logistic regression requires the dependent variable to have just two states, which are conventionally denoted 0 and 1. In many practical situations, however, we may be interested in exploring the influence of a set of explanatory variables on a response variable that has more than two categories. Such categories might be either *ordered* (as, for example, in a medical study on the effects of various factors influencing the 'level of pain', where the categories might be 'none', 'mild', 'moderate' and 'severe') or *unordered* (i.e. *nominal* as, for example, in a study investigating the effects of levels of pollutants on the incidences of several different types of disease). Generalisations of ordinary logistic regression are now widely available for both of these situations. In *ordinal logistic regression*, the logistic regression model is fitted to the *cumulative* logits, that is, logit(j) = $\log_e [\Pr(y \leq j)/\Pr(y > j)]$ where j denotes an ordered category; this model is commonly known as the *proportional odds model*. In *nominal logistic regression*, one of the categories (k say) is taken as a reference and the logistic regression is fitted to $\log_e(p_j/p_k)$ for all the other categories j; this model is commonly known as the *multinomial logit model*. While the computer output for these extensions is necessarily greater than that for ordinary logistic regressions, the procedures used in obtaining it are essentially the same as before.

Non-independence

This brings us to the last category of failures of assumptions in the fitting of linear models. Suppose that we have detected serial correlation in our set of data, and now wish to take account of it in our subsequent regression calculations. To discuss the possibilities, let us focus on the case of simple linear regression. In this case, the values y_i of the response variable are linked to the values x_i of the explanatory variable by the model $y_i = a + bx_i + \varepsilon_i$, where the departures ε_i come from a normal distribution with mean zero and variance σ^2 but are no longer mutually independent. So there is now in principle a full matrix of covariances between all possible pairs of departures and, consequently, between all possible pairs of y_i. We saw above how we could extend ordinary least squares to weighted least squares in order to accommodate unequal variances between observations. There is a generalisation of weighted least squares, not unexpectedly called *generalised least squares*, which will allow for having different variances and non-zero covariances when deriving parameter estimates, and this is one possible approach in the present case.

However, there are several drawbacks with the use of generalised least squares. One is that there are effectively many parameters (i.e. variances and covariances) that have to be estimated by sample quantities, and simply increasing the number of data values is no help because this introduces yet more parameters. A second one is that generalised least squares is often *too* general, in that it will react to every small sample perturbation or vagary of sampling—what is usually termed *overfitting* of the model, which makes the future predictions from this model even more hazardous than usual. A suitable compromise between ignoring the dependencies and overcompensating for them can be struck, as is usual in statistical analysis, by formulating a simple model that will adequately describe the correlation structure to within sampling variability. The parameters of this model can then be estimated, and the estimated model used in the regression analysis.

Since we have stressed that the lack of independence generally manifests itself as a serial correlation, we look for models that will reflect this type of dependence. Several classes of models have been developed for such structure in temporal and spatial studies, so these should be potentially useful here as well. The different classes of model are known respectively as *autoregressive* (AR), *moving average* (MA) and mixed *autoregressive/moving average* (ARMA) models, and each of these types of models can be of different *orders*, depending on how many terms are present. However, we will not confuse matters by describing each of these types. While in principle we could use any of these models to improve our regression procedure, it is usually sufficient to employ the simplest of them. This is the autoregressive process of order 1, denoted AR(1), and it is the recommended model because:

(i) it has been found adequate empirically over a wide range of situations;

(ii) often the reason for correlated ε_i is that some important variable z has been left out of the postulated relationship, and if z is slowly changing with time then the set of ε_i will look like an AR(1) process.

Thus the regression analysis can be improved by assuming that ε_i follows an AR(1) process rather than being independent (but note that it is probably better to find the 'missing' variable z and put it into an ordinary regression equation!).

The AR(1) assumption links the departure term ε_i to its predecessor in the serial order

$$\varepsilon_i = \alpha\varepsilon_{i-1} + v_i,$$

where α is an unknown parameter and v_i are values drawn independently from a normal distribution with mean zero and variance σ^2 (i.e. like departures in the ordinary least squares situation). Our simple regression model for dependent observations thus states that $y_i = a + bx_i + \varepsilon_i$, but with ε_i satisfying the above model rather than being independent.

A little bit of mathematics readily establishes that we can reduce this model to the form

$$y_i' = c + bx_i' + v_i,$$

where $y_i' = y_i - \alpha y_{i-1}$, $x_i' = x_i - \alpha x_{i-1}$, $c = (1 - \alpha)a$ and v_i is as above. So if we modify our y_i values by starting with the last one (y_n) and subtracting from each one α times the previous one, and then modify our x_i values in the same way, we can conduct the standard regression analysis on these modified variables because now the departures v_i satisfy the necessary assumptions. (Note, however, that we have to lose the original first pair of values x_1, y_1 because there is no preceding pair to subtract from them.) This standard regression analysis gives us 'correct' estimates of c and b. The latter is one of the estimates we wanted in the first place (the coefficient of x), while the other one we wanted can be obtained from the fact that $a = c/(1 - \alpha)$. However, α is generally unknown and so must be estimated. Various possibilities exist, but probably the easiest method is to estimate it by the serial correlation of the residuals used in the original check on validity of the independence assumption in the model.

This procedure extends in the obvious way to multiple regression situations. We simply have to conduct the above modification on *all* the explanatory variables in turn, and note that when we conduct the multiple regression on the modified variables then the regression coefficients do not need any further adjustment; only the regression constant is adjusted, exactly as above. Note also that if departures are correlated then this fact can be used to advantage in the prediction of future values, but we do not give technical details here.

Tailpiece

The scope for generalising the linear model is extremely wide. We have considered the more obvious generalisations under several generic headings, but because of the technical complications that are encountered in these generalisations we have been unable to give more than a very superficial account of each one. The reader should not underestimate the amount of work that has been done in these areas: whole text books have been written about the material touched upon in each section above, and such text books need to be consulted if a mastery of this material is sought. This chapter is perhaps just a glimpse through a doorway of the riches that lie beyond it.

8   Association between Variables

Introduction

Many scientific and social investigations are exploratory, at least in their early stages. The investigator may well have an overall objective or hypothesis in mind, but may not be entirely sure as to the nature of the samples to be collected or which variables should be measured on these samples. So a frequent approach is to measure as many variables as possible while the investigation is under way, and then to rely on statistical analysis to sift through the wealth of information that has been collected. The hope is that this analysis will elucidate any underlying data features that may be interesting; will highlight which of the variables may be important for future study and which of them can safely be ignored; and will produce a simplified picture that will help the researcher to formulate future hypotheses and future investigations to follow up. A collection of statistical techniques has been built up over the years in order to help with such analyses, and since they all relate to the joint analysis of more than one variable, this branch of statistics is usually termed *multivariate analysis*.

Note that in Chapters 6 and 7 we have already looked at some methods which can be employed when potentially many variables have been measured, but these methods all have the aim of either predicting or explaining the response of one of the variables, y say, by the measurements made on a series of other explanatory variables $x_1, x_2, \ldots, x_k$. Thus our interest in such cases is focused primarily on y, which therefore has a special status, and the other variables are subsidiary to this end. In multivariate analysis, however, there is no single variable that has special status, and our interest is spread evenly across the whole collection of variables. In recent years, the rapid development of automatic data recording instruments has meant that very large data bases can be built up easily, often involving the values of hundreds or even thousands of variables for each of perhaps tens of thousands of individuals, and methods of searching these data bases for recognisable patterns are constantly being sought. This is where multivariate methods are especially valuable, although when applied in such a context they are now more often referred to under the heading of *data mining*.

The central feature of multivariate data, and the aspect that leads to problems of interpretation and analysis, is the fact that the variables are associated because they have all been measured on each individual in the sample. Intrinsic properties of an individual will tend to affect, to a greater or lesser extent, all the measurements made on it and so a sample of individuals will give rise to associations between these measurements. For example, the size of an individual will be reflected in both its height and its weight, so a sample of individuals of different sizes will give rise to an association between these two measurements; a pupil's intelligence will affect his or her score on tests in a range of subjects, so a sample of individuals of differing intelligence levels will give rise to associations between all sets of scores; and so on. It is these associations that cause the complexities in the data structure, and they need to be first measured and then allowed for if we are to get to grips with what the data are telling us. So in this chapter we focus on the measurement of association, its interpretation and its possible explanation. Having examined association in detail in this way, we then turn in the next chapter to the exploitation of association between variables when exploring, simplifying and interpreting the data structure of the individuals in multivariate samples.

Measuring and testing association

As has already been stressed in Chapter 6, the measurement of association between two variables depends on whether they are quantitative or qualitative, so we consider each of these types separately.

Quantitative variables

Such variables have already been considered in Chapter 6, where we have seen that an appropriate measure of association between two quantitative variables is their correlation coefficient. For the three example pairs of variables in that chapter we found the correlations to be -0.177, -0.929 and 0.839 respectively, and we saw that these values accorded with the interpretation that the larger the absolute value of the correlation, the closer do the points lie to a straight line on the scatter diagram of the two variables.

However, the values are of course obtained from sample data (and in all three examples the samples are small), so a researcher might wish to make an inference in the usual way about the population value of the correlation coefficient in a particular case. One way would be to find a confidence interval for this 'true' value. Such an interval is easily constructed but it relies on technicalities regarding the sampling distribution of the correlation coefficient, so we do not give the details here; they can be found in most textbooks, and charts for constructing confidence intervals are also given in books of statistical tables. An alternative procedure would be to test a hypothesis about the value of the population correlation coefficient, and the generic null hypothesis is that the population correlation is zero against the general alternative that it is non-zero. There are tables from which this test can be carried out directly for a wide range of sample sizes, and most computer packages will print out a p-value for this test alongside a calculated correlation coefficient. For example, with a sample of 10 pairs of observations a correlation coefficient is significant at the 5% level (i.e. $p < 0.05$) whenever its absolute value exceeds 0.632, so the first correlation coefficient above is not significant while the other two are significant at this level.

However, there are several cautionary comments to be made about such a significance test, arising from the fact that the critical value for determining significance at a particular level is highly dependent on sample size. For example, if the sample size above were 5 instead of 10 then the critical value would be 0.878, for a sample of size 20 it would be 0.444, and for a sample of size 50 it would be 0.279. Clearly, therefore, a seemingly large correlation may be spurious if the sample size is small. Conversely, a seemingly small correlation may nonetheless be significant if the sample size is large. However, it should also be borne firmly in mind that rejection of the null hypothesis simply tells us that the population value is *not zero*, and tells us nothing about either its *actual* value or its practical importance. For example, if the sample size is large enough then it is possible to reject the null hypothesis when the true correlation is as low as 0.05, say, which is not a value of any practical interest!

If we have a multivariate data set, then there will possibly be many variables and a correlation coefficient can be calculated between every pair. With k variables there will be $k(k-1)/2$ such pairs. These correlation values are most conveniently arranged in a block of k rows and k columns, so if the variables are $x_1, x_2, \ldots, x_k$ then the correlation between variables x_i and x_j is placed in row i and column j of this block; this is known as the *correlation matrix* of the data. Of course, the correlation between x_i and itself is 1.0 (as a scatter diagram of x_i against itself will have points lying exactly on a 45° line), so the top left to bottom right diagonal of this matrix contains 1.0 in all positions. Also, the correlation between x_i and x_j is the same as the correlation between x_j and x_i, so the values above this diagonal are reflections of those below the diagonal (and the matrix is then said to be *symmetric*). For example, in an educational study, measurements were taken on the intelligence quotient (IQ) score, the weight in kilograms and the age in months of each of 45 school children, and the IQ/weight, IQ/age and weight/age correlations were 0.616, 0.827

and 0.732, respectively. These correlations can thus be exhibited in the matrix

$$\begin{array}{c} \text{IQ} \\ \text{Weight} \\ \text{Age} \end{array} \begin{pmatrix} 1.0 & 0.616 & 0.827 \\ 0.616 & 1.0 & 0.732 \\ 0.827 & 0.732 & 1.0 \end{pmatrix},$$

$$\text{IQ} \quad \text{Weight} \quad \text{Age}$$

which has all the properties given above, and where the labelling in the borders is often included for ease of identification of the matrix entries.

Whenever such a matrix is produced by a statistical package there will typically be p-values for the test of zero population correlation printed alongside each value in the matrix. The temptation for the researcher is to go through the matrix identifying 'significant' correlations, with the aim of highlighting groups of 'associated' and groups of 'unassociated' variables. By this means, the researcher often hopes to identify some of the most 'important' variables in the study. However, there are several dangers in such a strategy. First, a significance level of, say, 5% means that 5% of all tests in which the null hypothesis is true will give an incorrectly significant outcome, so if there are many entries in the matrix then there will be many such 'false significances'. For example, if there are $k = 20$ variables then there will be $k(k - 1)/2 = 190$ different correlations and therefore we should expect around 9 or 10 such false significances—and we will not of course know which of our significances are genuine and which are false. Second, since we are calculating the correlations between all pairs of variables, the entries in the correlation matrix are not independent of each other; indeed, there is considerable correlation between the various correlation values! So attempting to interpret them all independently is very hazardous. We will describe rather better approaches to the interpretation of many linked correlations below, but first let us consider association between qualitative variables.

Qualitative variables

We have seen in earlier chapters that discrete variables are typically represented by the possible values of the variable plus frequencies of occurrence of each value in the sample. More generally, qualitative variables will have a number of different possible *categories* (e.g. the qualitative variable 'eye colour' will have categories 'blue', 'brown', 'green', 'black' and 'grey') and a sample will be represented by counts of members in the different categories. Analogously to the scatter diagram for quantitative data, we would like to see the joint action of two qualitative variables and this is best provided by *a contingency table*. In such a table the rows are the categories

Table 8.1. Contingency table showing cross-classification of 837 workers by education and political affiliation

Education	Party voted for			Total
	Conservative	Labour	Liberal Democrat	
University educated	80	45	80	205
Not university educated	387	84	161	632
Total	467	129	241	837

Source: Reprinted from Diamond and Jefferies (2001) by permission of Sage Publications Ltd.

of one variable, the columns are the categories of the other variable, the cells give the number of sample members jointly falling into the two relevant categories, and the margins give the totals in each of the rows or columns. For example, Diamond and Jefferies (2001) report the data in Table 8.1 from the 1987 British Election Study in which 837 randomly sampled salaried workers were classified according to whether or not they received a university education, and which of the three main political parties they voted for.

Thus there were 80 Conservative voters who had a university education, 632 individuals in the sample who did not have a university education, 241 Liberal Democrat voters, and so on. A question of interest is whether voting behaviour is associated with university education. If it is *not* then the two attributes are independent (i.e. one does not influence the other), and roughly the same proportions should be found in each of the three voting cells for the 'university educated' row of the table as for the 'not university educated' row and the 'total' row. This accords with the technical definition of independence given in Chapter 1 on probability, viz. that the marginal probabilities of randomly chosen individuals voting Conservative, Labour or Liberal Democrat respectively should equal their corresponding conditional probabilities given education level. Any departure from independence is then deemed to indicate association, and the greater the departure the stronger is this association.

Looking at the numbers in the table, there certainly seems to be an indication of association as the proportion of Conservative voters is greater for the 'not university educated' group than for the 'university educated' one while the other two proportions show the reverse pattern. However, how strong is this association and is it 'real' or just attributable to chance sample fluctuation?

To measure departure from independence, we compare the observed values in the table with the values we would expect to have in each cell of the table if the variables were genuinely independent. To see how the latter are calculated, consider the cell of the table corresponding to the joint event 'Conservative voter' and 'university educated'. From the values in the table, the estimated probability that a randomly chosen person is university educated is $\frac{205}{837}$ while the estimated probability that a randomly chosen person votes Conservative is $\frac{467}{837}$. Then *if* voting pattern *is* independent of education, the probability that a randomly chosen person is both university educated *and* votes Conservative is just the product of these two fractions, namely $\frac{205}{837} \times \frac{467}{837}$. Thus out of 837 people, we would expect $837 \times \frac{205}{837} \times \frac{467}{837} = \frac{(205 \times 467)}{837} = 114.38$ (to two decimal places) people to both have had a university education and to vote Conservative.

We can follow this argument through for every cell and obtain the same pattern, so in general if the variable governing rows is independent of the one governing columns then the expected value in row i and column j of the table is

$$(\text{Total in row } i) \times (\text{Total in column } j) \div (\text{Grand total})$$

Completing these calculations for every cell of the voting behaviour data gives the expected values in Table 8.2.

Two things ought to be noted when performing these calculations. Although the observed values are all counts and hence integers, the expected values should be calculated accurately (to at least two decimal places) and *not* rounded to integers. This is because we use a continuous distribution to approximate the sampling distribution of the test statistic specified below. Second, the margins of the table of expected values should always agree with those of the observed values, as the sampling process fixes these margins; this provides a useful check on the calculations.

Let us denote the observed value in row i and column j by O_{ij} and the expected value in row i and column j by E_{ij}. If the two classifying variables are independent, all the E_{ij} should be 'close to' their O_{ij} counterparts. Looking at the two tables above, it is evident that there are some considerable discrepan-

Table 8.2. Expected values in the cells of the contingency table of Table 8.1, assuming independence between education and political affiliation

Education	Party voted for			Total
	Conservative	Labour	Liberal Democrat	
University educated	114.38	31.59	59.03	205
Not university educated	352.62	97.41	181.97	632
Total	467	129	241	837

Source: Reprinted from Diamond and Jefferies (2001) by permission of Sage Publications Ltd.

cies so there seems to be evidence of association between education and voting behaviour. But how strong is this evidence? To answer this question we must somehow combine all the discrepancies $O_{ij} - E_{ij}$ into a single measure, and this is done by using the Pearson goodness-of-fit statistic that we have already met in Chapter 4, namely by squaring each discrepancy, dividing it by E_{ij} and adding up all the resulting values to give a single quantity which we denote by X. This sequence of steps can be encapsulated in the single formula

$$X = \sum_{i,j} \frac{(O_{ij} - E_{ij})^2}{E_{ij}}.$$

These steps are summarised in Table 8.3 for the voting behaviour data.

Following through the logic behind the calculation, if the two classifying variables are independent then X should be close to zero, but if they are associated then it should be 'large'. However, as always we have to account for sampling fluctuations and theory tells us that if there is no association then the sampling distribution of X is the chi-squared distribution with $d =$ (number of rows $-$ 1) $\times$ (number of columns $-$ 1) degrees of freedom. On the other hand, if there *is* association then X will be 'large' relative to this distribution. The formal procedure we adopt is thus to test the null hypothesis H_0: there is no association between the variables, against the alternative H_1: the variables are associated, by seeing if the calculated value of X lies in the upper tail area of the chi-squared distribution on d degrees of freedom. If it does then it is deemed to be 'large' so we reject H_0 in favour of H_1 and conclude that the variables are associated, but if it does not then we do not reject H_0 and conclude that the variables are independent. The tail area for appropriate significance levels is found from the table of percentage points of the chi-squared distribution.

For the voting behaviour data, we have $d = (2 - 1) \times (3 - 1) = 1 \times 2 = 2$. So for a 5% significance level, the tail area contains all values greater than 5.991; for 1%, all values greater than 9.210 and for 0.1%, all values greater than 13.815. The calculated value of 31.09 for X lies in all these tail

Table 8.3. Steps in the calculation of X for the voting behaviour data

O_{ij}	E_{ij}	$O_{ij} - E_{ij}$	$(O_{ij} - E_{ij})^2$	$(O_{ij} - E_{ij})^2 / E_{ij}$
80	114.38	-34.38	1181.98	10.33
45	31.59	13.41	179.83	5.69
80	59.03	20.97	439.74	7.45
387	352.62	34.38	1181.98	3.35
84	97.41	-13.41	179.83	1.85
161	181.97	-20.97	439.74	2.42
				$X = 31.09$

Table 8.4. Survival of infants according to the
amount of antenatal care received

Care received	Number died	Number survived	Total
Less	20	373	393
More	6	316	322
Total	26	689	715

areas, so there is very strong evidence of association between voting behaviour and education level.

This test procedure is commonly termed the 'chi-squared test of association'. It works for any size of table, but needs all E_{ij} to be at least 3 for validity. If any E_{ij} are less than 3, then the table must be reduced by combining appropriate rows or columns and the test must be recalculated for the new table.

Explaining associations: partial correlations

Establishing the presence of association between variables is undoubtedly important, but perhaps it is even more important to understand why this association has arisen. Indeed, much research work in social and behavioural sciences in particular is aimed at finding good explanations for observed relationships. One very frequent cause of an observed association between two variables is their mutual association with a third variable, which exhibits a range of values in the sample under consideration. If the association between the first two variables disappears when the third variable is held constant, then we can say that this association has been 'explained' by this third variable.

This idea is most easily illustrated by a simple example with qualitative variables, given in *An Introduction to Latent Variable Models* by B.S. Everitt (1984). The data in Table 8.4 show the survival of infants according to the amount of antenatal care they received (as assessed by an outside paediatrician into the two categories 'less' and 'more'). Applying the chi-squared test of association to these numbers, we find that $X = 5.26$ on $d = 1$ degree of freedom. Hence the significance level of the test is less than 5%, and it appears that the survival of infants is indeed associated with the amount of antenatal care they receive—an entirely reasonable conclusion.

However, it transpires that these infants were born in two different hospitals, so we can introduce 'Hospital' as an extra qualitative variable and disaggregate the values in Table 8.4 according to this third variable. We find that for Hospital A the incidences were

Care received	Number died	Number survived	Total
Less	3	176	179
More	4	293	297
Total	7	469	476

while for Hospital B they were

Care received	Number died	Number survived	Total
Less	17	197	214
More	2	23	25
Total	19	220	239

Applying the chi-squared test to each of these tables in turn, we find that X is very close to zero for each one and hence there is no association between survival rate and care for the separate hospitals. Thus an apparent association between survival rate and care for data in which a third variable (Hospital) has a range of values disappears when this third variable is held at a constant value. On examining the data more closely, we see the reason why this has happened. The two hospitals have very different rates of births (about twice as many in Hospital A as in Hospital B), and very different (apparent) policies on amount of care (Hospital A accords 'more' care to a large proportion of infants, while Hospital B accords it to a much smaller proportion). Whatever the reasons for such variations in rates and policies, the effect on the data is such that when the numbers for the two hospitals are combined, a lack of association in each separately is transformed into a significant association for the aggregated data.

Of course, some of the numbers in this example are very small, and one might question the validity of chi-squared tests in such circumstances, but the underlying message should be clear: do not take an observed association at face value, but check to see whether disaggregating along the lines above will account for it. Even if disaggregating according to a single variable does not do so, it is possible that disaggregating according to the joint categories of two or more variables will remove the association.

The same idea applies to quantitative data, but seeing how it works is a little less obvious. Essentially, suppose that we have three variables x, y and z, say, and that each of x and y are (separately) highly positively correlated with z. Then scatter diagrams of x versus z and y versus z will both have points clustering about straight lines, such that as z increases then x and y both separately increase. Now suppose that the sample we have gathered contains a good range of values of z. Because of the relationships above, low values of z will tend to be linked to low values of both x and y, while high values of z will be associated with high values of both x and y. Consequently, a scatter diagram of x against y will show a distinctly linear trend among the points, and hence there will be a substantial correlation coefficient *whether or not x and y are 'genuinely' correlated*. In order to determine whether there is a 'real' relationship between x and y, rather than it being an artefact of their joint relationships with z alongside a range of z values in the sample, we would need to do something like we did with the qualitative data, namely examine the relationships at *fixed* values of z. If the correlation coefficient remains high then there is a 'real' relationship between x and y, but if the correlation falls to near zero then the apparent relationship is just an artefact.

Unfortunately, with continuous data we are not usually in a position where we can disaggregate the sample into separate groups at each distinct value of z in the way that we could with qualitative data. This is because the values of z occur in a continuum, and in many situations we do not even have any repeated values of z, let alone groups at just a few values. So now we need to *estimate* what the correlation between x and y *would be* if z were held fixed at a constant value, and this is done by calculating the *partial correlation* between x and y for fixed z. If we denote the ordinary correlation between x and y by r_{xy}, then we usually write $r_{xy \cdot z}$ for the partial correlation between x and y for fixed z. This partial correlation is an estimate of how much correlation remains between x and y after their separate relationships with z have been eliminated. Since correlation is a linear concept, the residuals from the linear regression of x on z indicate what is 'left over' after the relationship between x and z has been allowed for, and similarly the residuals from the linear regression of y on z are what is 'left over' after the relationship between y and z has been eliminated. Clearly, therefore, $r_{xy \cdot z}$ is just the correlation between these two sets of residuals. However, an alternative (and quicker) way of calculating it is by subtracting the product of the correlation between x and z and the correlation between y and z (i.e. $r_{xz} \times r_{yz}$) from r_{xy}, and normalising the result so that it lies in the range -1 to $+1$ by dividing it by $\sqrt{(1 - r_{xz}^2) \times (1 - r_{yz}^2)}$.

As an example, return to the above study regarding the IQ of school children and denote IQ, weight and age by x, y and z, respectively. We have a high positive correlation between IQ and weight, $r_{xy} = 0.616$, so do we deduce that we increase children's brainpower by feeding them lots of chocolate bars? Certainly not, because the sample contained children of different ages (variable z), and the respective correlations between age and each of the other variables were $r_{xz} = 0.827$, $r_{yz} = 0.732$. Thus

$$r_{xy} - r_{xz} \times r_{yz} = 0.616 - 0.827 \times 0.732 = 0.616 - 0.605 = 0.011.$$

Also,

$$\sqrt{(1 - r_{xz}^2) \times (1 - r_{yz}^2)} = \sqrt{(1 - 0.827^2) \times (1 - 0.732^2)}$$
$$= \sqrt{0.316 \times 0.464} = 0.383,$$

so that finally $r_{xy.z} = 0.011 \div 0.383 = 0.029$. Thus, if age were to be kept constant (i.e. for pupils of the same age), the correlation between weight and IQ would shrink to near zero. Hence the original correlation is an artefact of the joint relationships between weight and IQ with age, allied to the range of ages in the sample. Note, however, that if we adjust the correlation of 0.827 between IQ and age on fixing weight, we find that there is only a very modest decrease to $r_{xy.z} = 0.701$, so that the IQ/age correlation is not 'explainable' by weight.

As with qualitative variables, the correlation between two variables can be corrected for any number of other variables if it is suspected that the original relationship is an artefact of joint relationships with them. Although extensions to the above formulae for direct correction are possible, the process becomes increasingly complicated if more than one or two variables are to be kept fixed. The simplest general approach is therefore to return to the idea of residuals from regressions as indicators of what is 'left over' after particular relationships have been allowed for, and to correlate the residuals from multiple regressions in which each of the two primary variables is regressed on the set of variables that are to be kept fixed. Also, formal tests of hypotheses can be conducted on partial correlations by using the same sets of tables as for ordinary correlations, the only adjustment being that the number of sample points has to be reduced by the number of variables kept fixed in the partial correlation.

Explaining associations: latent variable models

Now suppose that we have observed a set of k quantitative variables $x_1, x_2, \ldots, x_k$ and we calculate the correlations between all pairs of these variables. Researchers in social and behavioural sciences often wish to 'explain' *all* these correlations in terms of various common features. It is conceivable that we have available, or we subsequently collect, a further set of variables $z_1, z_2, \ldots, z_q$ that could potentially be used for such an explanation. Then the following systematic procedure based on the foregoing ideas could be employed.

Write r_{ij} for the correlation between x_i and x_j, and let all the correlations be written in the correlation matrix $\mathbf{R}$. We first search through the further variables and identify the one, z_a say, for which the partial correlations $r_{ij.z_a}$ are as small as possible for all i, j pairs. If none of these partial correlations is significantly different from zero, then we can say (in terms of the above interpretation) that z_a has 'explained' all the correlations in $\mathbf{R}$. But if any of the partial correlations are significantly different from zero then there is still some correlation to be explained, so we now search for a second further variable, z_b say, for which the partial correlations $r_{ij.z_a z_b}$ are as small as possible for all i, j pairs. Once again we test all the partial correlations, and if any are significantly different from zero we continue the search. Eventually (we hope) there is a set $z_1, z_2, \ldots, z_m$ of these further variables for which all the partial correlations $r_{ij \cdot z_1 z_2 \ldots z_m}$ are not significantly different from zero, in which case $z_1, z_2, \ldots, z_m$ explain $x_1, x_2, \ldots, x_k$ in the sense that all correlations between the x variables vanish when the z variables are held fixed at constant values. In such a case, we say that the x variables are *conditionally independent* (or *locally independent*) given the z variables, and this property is equivalent to saying that each x can be modelled as a linear combination of all the z variables, plus a departure term that is independent of all the other departure terms. (This follows from the earlier discussion, because if the x variables can be written in this way then the residuals from a multiple regression of each x variable on all the z variables simply estimate the model departure terms, and mutually independent departures mean that every pair of residuals should be uncorrelated.)

Unfortunately, it is very rare in practice that such a set of further observable (or *manifest*) variables either exist or can be found, so it seems that there is an impasse and that explaining all the correlations in the matrix **R** is an impossible task. The ingenious solution, first propounded at the turn of the twentieth century, is to assume that the z variables *do* exist, *but they can never be observed*. Thus they are termed *latent* variables, by contrast with the manifest variables that *can* be observed. Such unobservable variables would typically represent constructs like 'social class', 'public opinion', 'extroversion/introversion', and so on. This has led to the development of *latent variable models* for the analysis of association between variables, and there are different types of such models depending on the form (qualitative/ categorical or quantitative) that each x and each z are assumed to be. If the x and z variables are all qualitative then we have a *latent class model*; x quantitative but z qualitative gives a *latent profile model*; x qualitative but z quantitative leads to a *latent trait model*; while if x and z are all quantitative then we have a *factor analysis model*. The last named is the oldest model and the most common one in practical applications, so the term 'factor' is now used generally instead of 'latent variable'. The bulk of our discussion will therefore focus on the factor analysis model, with some brief comments at the end about the other models. Given the somewhat amorphous nature of latent variables, these techniques find their main uses in social and behavioural studies, with relatively few instances of their application in the more exact sciences. But that of course is not to rule them out entirely, and realistic examples of their utility in such studies do occur from time to time.

Historically, latent variable modelling dates from the studies of human abilities conducted by Charles Spearman and others at the end of the nineteenth, beginning of the twentieth centuries. Early work concentrated on the simplest factor analysis models involving just a single factor. Louis Leon Thurstone and colleagues then developed more general models in the

1920s and 1930s, but poor estimation techniques bedevilled the area and hampered its general acceptance. Some improvements came with the maximum likelihood approach pioneered by D.N. Lawley in the 1940s, but this approach still encountered computational problems and these were not fully resolved until the work of Karl Jöreskog and others in the 1970s. In the meantime, other types of latent variable models were being developed from the 1950s onwards, with commensurate advancement in computational effectiveness in the 1970s and 1980s, to bring us to the present-day position in which comprehensive software for fitting a whole range of models is widely available through software packages such as LISREL and EQS.

Factor analysis methodology

Following the above discussion, let us assume that we have observed a set of k quantitative variables $x_1, x_2, \ldots, x_k$ on a number of individuals, and that we wish to explain the resulting correlations between all pairs of these variables. For the purposes of the development, we will assume that these variables have all been mean-centred (i.e. each individual value is expressed as a residual by subtracting the relevant variable mean). It is often useful to assume also that they have been standardised (i.e. each residual is divided by the variable's standard deviation), but this is not essential.

At the heart of factor analysis is the *factor model*, which postulates that each manifest variable is made up of a linear combination of a set of m *common factors* $z_1, z_2, \ldots, z_m$ plus a *specific variate*. Thus we can write the ith manifest variable x_i as

$$x_i = \lambda_{i1}z_1 + \lambda_{i2}z_2 + \cdots + \lambda_{im}z_m + \varepsilon_i,$$

where λ_{ij} is known as the *loading* of the jth factor z_j on the ith variable x_i and ε_i is the specific variate associated with x_i. In other words, this model says that x_i is in effect 'made-up' of contributions from each of the common factors $z_1, z_2, \ldots, z_m$, the 'amount' of z_j in this contribution being λ_{ij}, plus an extra amount ε_i that is specific to x_i and not 'shared' with any other x variable. The fact that every x contains contributions from all the zs (to varying amounts) is what produces the correlations between every pair of xs, but every x is also allowed to have its 'own' portion ε_i as otherwise the model would be far too restrictive.

We note that the above model bears superficial resemblance to the multiple regression model of Chapter 6, with the manifest variable x_i replacing the former response variable y, the factors $z_1, z_2, \ldots, z_m$ replacing the former explanatory variables, the loadings λ_{ij} replacing the former coefficients β_j, and the specific variate ε_i playing the role of the former departure term. However, the big difference between the two models is that the explanatory variables in multiple regression are observed and so carry values for each individual in the sample, but the factors in the factor model are unobserved and so do not have values for each individual. In fact, *none* of the quantities on the right-hand side of the above equation is observed, the only sample values being attached to x_i on the left-hand side. Thus, whereas in multiple regression we only needed to estimate the regression coefficient β_j, in factor analysis we now need to estimate the values of each factor (i.e. the *factor scores*) for each individual as well as the factor loadings λ_{ij}.

It may seem to be an extremely tall order to be asked to estimate so many quantities from what is a fairly restricted set of sample observations, but it turns out to be entirely possible. However, in order for it to be possible, and for the resulting estimates to be unique, we need to make several distributional assumptions about the common factors and the specific variates, and also to impose some constraints on the number of factors and on the loadings in the model.

The model assumptions are:

- The common factors are *mutually independent* random variables, each having mean equal to zero and variance equal to one.
- The specific variates are independent of each other, and of all the common factors. They all have mean equal to zero, but their variances are not constrained to be equal so for generality we let the variance of ε_i be σ_i^2.
- For maximum likelihood estimation (see below), the common factors and specific variates also have to be normally distributed.

The constraint on the number m of factors in the model is that $k + m$ must not exceed $(k - m)^2$ (where k is the number of manifest variables). So, for example, if we have three manifest variables then there can at most be just one common factor in the model. To include two common factors we need at least five manifest variables, for three common factors we need six manifest variables, for four common factors we need at least eight manifest variables, and so on. The constraints on the loadings are quite complicated technical ones, but are imposed purely to ensure that the same unique set of loadings is obtained in any situation by any user. They are automatically incorporated in all factor analysis software routines, and therefore need not concern us here.

The mathematical implications of the model are that the variance of x_i is

$$\lambda_{i1}^2 + \lambda_{i2}^2 + \cdots + \lambda_{im}^2 + \sigma_i^2,$$

while the covariance between x_i and x_j is

$$\lambda_{i1}\lambda_{j1} + \lambda_{i2}\lambda_{j2} + \cdots + \lambda_{im}\lambda_{jm}.$$

The quantity $\lambda_{i1}^2 + \lambda_{i2}^2 + \cdots + \lambda_{im}^2$ that forms the first part of the variance is known as the *communality* of x_i, because it is the part of its variance attributable to the common factors, while σ_i^2 is the part specific to x_i. Thus if the communality of a manifest variable forms a large proportion of its variance, then the common factors play a major role in explaining that variable. Note also that if the manifest variables have all been standardised then the covariance between any two of them is equal to the correlation between their unstandardised versions.

Thus, the model provides theoretical values for the variances of all the manifest variables and for the covariances or correlations between every pair of manifest variables. The sample data provide the corresponding sample values for all these variances and covariances/correlations. So if there are indeed m common factors that explain all the associations between the variables, the sample values should all be 'close to' the theoretical values predicted from the model. The purpose of factor analysis is therefore to find the value of m, and values of all the loadings λ_{ij} and specific variances σ_i^2, such that the sample values are all 'as close as possible' to the corresponding theoretical values.

The problem with the phrase 'as close as possible' is that it can be interpreted in many different ways, so there are correspondingly many different methods of conducting factor analysis. Indeed, this was the problem in the historical development of the technique, as some of these methods became discredited and thus the technique was viewed with suspicion. However, as has been already mentioned, there are now various perfectly sound estimation methods, of which the ones using *least squares* and *maximum likelihood* are the two most popular. Both approaches are iterative ones, in other words the computer starts from an initial guess at the solution and then gradually improves it until it can do no better. Both also, unfortunately, rest on sophisticated mathematics for their justification, so we will content ourselves here with just a brief outline of the essential steps of each method.

Least squares makes no assumptions about distributional form, so no assumptions about normality of common factors or specific variates are required. The method requires initial guesses to be made at the set of specific variances σ_i^2, subtracts each guess from the sample variance of the appropriate manifest variable to form 'reduced variances' and then estimates all the loadings λ_{ij} using principal component analysis (see Chapter 9). This analysis is conducted on the matrix of variances and covariances if standardising the manifest variables is not appropriate, and on the correlation matrix if it is. The estimated loadings in turn provide estimates of all the communalities, which when subtracted from the sample variances provide improved estimates of σ_i^2. These improved estimates are then used in a repeat of the foregoing steps. The cycle continues until two successive computations yield the same σ_i^2 values (to a given number of decimal places), when it is terminated and the current values of λ_{ij} and σ_i^2 are taken as the solution. This approach is often called the 'principal factoring' method.

By contrast, maximum likelihood *does* require distributional assumptions, so the user must be satisfied that the data approximate to a normal distribution before this method is invoked. If normality is reasonable, and if the factor model is correct, then it is possible to write down a (fairly complicated) mathematical expression for the probability of the sample data, where the loadings and specific variances are included in this expression. The computer then finds the values of these quantities that make the probability of the sample as large as it can be. These values are the maximum likelihood estimates. The process is again iterative, finding values of specific variances and then loadings in turn until no further improvement is obtained.

However, both the least squares and the maximum likelihood methods require the number m of factors in the model to be prespecified, so how do we decide on the appropriate value? The most common approach is to proceed sequentially, fitting the simplest model (i.e. just one factor) first and seeing if the model fits satisfactorily. If it does then we stop, but if it does not then we fit the next simplest model (i.e. two factors) and see if this fits satisfactorily or not. We proceed in this way, gradually fitting more factors until the fit becomes satisfactory. At this point we are satisfied that we have found the simplest model that will adequately account for all our observed associations. But all this begs the question: How do we assess the fit of a given model to the available data? The real power of maximum likelihood comes to the fore here, because if we can assume normality of data then we can conduct a significance test of adequacy of the model against the alternative that a more complicated one is needed. This is a chi-squared test, which is available on all computer packages that carry the maximum likelihood factor analysis software, so it can be conducted objectively at each stage of the model fitting procedure. Rejection of the null hypothesis implies that a more complicated model is needed, so we continue fitting extra factors until the test becomes non-significant. However, if we are using least squares without assumption of normality, then we must apply a subjective check of the model. Various possibilities exist, but probably the best one is to find the residual variances and covariances (i.e. the differences between the sample values and their predictions from the fitted model). If these residuals are all 'small' then the model fits well, but if some are not small then we need to go on to the next model in the sequence. Unfortunately, no hard and fast thresholds of 'smallness' can be offered, so this is very much a subjective procedure.

There is one other very powerful property of maximum likelihood that gives it the edge over least squares for factor analysis. This is the property of *invariance*, which in factor analysis means that the same results and conclusions are obtained whether the data are standardised or not, that is, whether the covariances or the correlations are used for the analysis. Unfortunately, least squares leads to different outcomes in the two cases, so a careful decision must first be made as to whether or not to standardise before the analysis is undertaken.

Having found a model that fits, the first task is usually to try and interpret the factors produced. This is done by identifying the 'important variables' associated with each factor (those whose loadings are not 'close to zero') and then seeing what their combination means in substantive terms. Once again this process involves considerable subjectivity, but the task becomes easier when the researchers are very familiar with the variables in a particular subject area. However, there is one major problem with factor analysis, and that is that the factor model is not unique: The mathematical form remains unchanged if a linear transformation is applied to the factors, providing that the inverse transformation is applied to the loadings. What this means is that there are many 'alternative versions' of the solution, all of which are equally acceptable. It is therefore worthwhile seeking out the solution that is the easiest one to interpret. This is usually reckoned to be the solution that has the *simplest structure*, and a simple structure is when each factor has most of its loadings near zero and only a few with 'substantial' values. If only a few variables are associated with each factor, then that makes the factors much easier to interpret.

Factor rotation is the process of seeking such alternative representations, and much effort has been expended by researchers into ways of obtaining optimal rotations automatically. Various such rotations are now available in statistical packages, including rotation to either *orthogonal* (i.e. uncorrelated) or *oblique* (i.e. correlated) factors. The former are usually preferred as being easier to handle, and of such rotations VARIMAX is the most popular one in practice. Use of this option in a package will generally produce the factor solution with simplest structure for interpretation.

Having found a well-fitting model, and interpreted the resulting factors (e.g. general intelligence, verbal ability, numerical ability for a set of examination marks), it is often of interest to estimate each sample member's 'score' on each factor. In fact, measuring a person's intelligence quotient (IQ) is just the process of estimating that person's score on the factor 'general intelligence'. However, this is again a non-trivial statistical estimation problem, which can be tackled in a variety of ways. Unsurprisingly, there are two popular approaches, one relying on least squares and one on conditional normal estimation. Both lead to a 'factor score matrix' which gives the values that each person's manifest variable values must be multiplied by to give that person's estimated factor scores, and printing such factor score matrices plus resultant factor scores are standard options in computer packages.

To illustrate some of these ideas, consider a set of data first described by Yule *et al.* (1969). The full data comprised marks (0–20) obtained by 150 children aged 4–6$\frac{1}{2}$ years on each of the following 10 subtests of the Wechsler Pre-school and Primary Scale of Intelligence (WPPSI): 1. Information; 2. Vocabulary; 3. Arithmetic; 4. Similarities; 5. Comprehension; 6. Animal House; 7. Picture Completion; 8. Mazes; 9. Geometric Design; 10. Block Design.

Table 8.5. Maximum likelihood factor loadings for the WPPSI data, for both unrotated and VARIMAX rotated solutions

Subtest	Factor 1 (Unrotated)	Factor 2 (Unrotated)	Factor 1 (VARIMAX)	Factor 2 (VARIMAX)
1. Information	0.79	−0.40	0.85	0.25
2. Vocabulary	0.83	−0.24	0.77	0.40
3. Arithmetic	0.74	−0.03	0.56	0.48
4. Similarities	0.59	−0.19	0.55	0.27
5. Comprehension	0.68	−0.25	0.66	0.28
6. Animal House	0.65	0.14	0.38	0.55
7. Picture Completion	0.64	0.23	0.31	0.61
8. Mazes	0.63	0.35	0.22	0.69
9. Geometric Design	0.56	0.05	0.37	0.43
10. Block Design	0.81	0.41	0.30	0.86

For a sample of 75 children, all correlations between the marks of pairs of subsets were positive and substantial (mostly between 0.4 and 0.7). This suggested that the 10 subtests could be replaced by relatively fewer uncorrelated factors, representing different facets of the children's abilities. On applying maximum likelihood factor analysis, the chi-squared test and the residual matrix both pointed to a two-factor model as an adequate explanation of the correlations. The resulting factor loadings are shown in Table 8.5, the first two columns giving the unrotated loadings and the last two columns giving the loadings after VARIMAX rotation.

In the unrotated solution all subtests have substantial (positive) loadings on the first factor, which can therefore be interpreted as 'general ability': a high score on this factor requires high marks on all subtests. By contrast, the second factor is bipolar with positive loadings on the spatial/pictorial subtests 6–10 and negative loadings on the verbal/numerical subtests 1–5. Children who have better spatial/pictorial marks than verbal/numerical ones will have a positive score on this factor, while those with the reverse pattern will have a negative score. However, bipolar factors are generally disliked, as they do not admit concise interpretations, so here a VARIMAX rotation is useful. We see from the last two columns of Table 8.5 that all negative loadings are now removed: factor 1 has high positive loadings for subtests 1–5 and much lower ones for subtests 6–10, while factor 2 shows the reverse pattern. Thus factor 1 is interpreted as verbal/numerical ability, and factor 2 as spatial/pictorial ability. We could now obtain the factor score matrix, and hence determine each child's scores on these two factors as estimates of the respective abilities.

Other latent variable models

The above methodology is applicable to quantitative manifest variables, in particular to ones consonant with a normal distribution if maximum likelihood is to be employed. So what can we do if the manifest variables are qualitative, binary or categorical? An early approach was to replace the starting point of correlations between every pair of manifest variables by analogous measures for non-quantitative data (such as tetrachoric or polychoric correlations), and then to carry on with the least squares approach given above. More recently, the implicit linear regression structure of the factor model has been replaced by an appropriate function from the generalised linear model formulation described in Chapter 7, and the estimation procedure has been adapted to take account of this extra complexity. Such extensions are known as latent trait models. However, if the manifest variables are categorical then it may be more appropriate to make the latent variable categorical as well, and thus move to latent class models.

The first point to note about such models is that, by contrast with factor models, there is little purpose in having more than one categorical latent variable. This is because we are usually only interested in the categories, that is, *classes*, that the individuals fall into rather than in interpreting different dimensions of these classes, and two categorical variables with a and b categories respectively can always be replaced by a single variable having $a \times b$ categories. This argument, of course, extends to any number of variables. So we focus on a single categorical latent variable having c categories or classes.

If we have several categorical manifest variables, the data can be displayed in a contingency table; the ones in Tables 8.1 and 8.4 are examples where the number of manifest variables is two, but drawing up contingency tables having a larger number of manifest variables presents no problems. Efficient display of such higher-order tables requires the categories to be arranged in a systematic fashion, but no new ideas are introduced. Returning to Table 8.4, we saw that the (significant) association between survival of infants and amount of antenatal care could be explained by disaggregating the infants according to the hospital in which they were born, because the association vanished within each hospital. Had we not been able to identify the infants on a third variable such as 'hospital', we could have tried to segregate them into two different groups in such a way that there was no association between survival and antenatal care for each group. This is essentially the basis of a latent class model: we try to find a number of classes into which to allocate the individuals, so that within each class the manifest variables are mutually independent. A purely computational approach would involve trying out all possible arrangements into the various classes, and choosing the one for which the within-class associations were smallest (in some overall sense). However, it is evident that such a search would usually founder because of the huge amount of computer time that would be needed for all but the smallest practical data sets. So the obvious alternative is to formulate a suitable probability model, and use the data to estimate the parameters of the model and make the assignment into classes.

Here we outline the formal approach, but do not consider any of the technical details of its implementation. Let us suppose that there are k manifest categorical variables $x_1, x_2, \ldots, x_k$, and that we are seeking a categorical latent variable z that has c classes. Furthermore, let us confine ourselves for the time being to the simplest possible case where each manifest variable has just two categories, which we can denote arbitrarily as 0 and 1. So the data of Table 8.4 are of this form, with c and k both equal to two. In order to formulate a probabilistic model for the data, we must now specify some probabilities. Let θ_{ij} be the probability that an individual in class j of z exhibits category 1 of the manifest variable x_i (so that $1 - \theta_{ij}$ is the probability that an individual in class j of z exhibits category 0 of x_i). Now, *since the latent classes obey conditional independence*, the probability that an individual in class j of z exhibits a particular pattern p of categories for *all* the manifest variables $x_1, x_2, \ldots, x_k$ is just given by the product of θ_{ij} or $1 - \theta_{ij}$ terms appropriate to the pattern p; call this probability $\phi(p \mid j)$, adopting this notation because it is strictly speaking a conditional probability (of the pattern p conditional on class j of z). We can obtain this probability for any pattern p and any class j in terms of the underlying parameters θ_{ij}. Now suppose that the classes j have (prior) probabilities α_j for $j = 1, \ldots, c$. Then, by the total probability theorem of Chapter 1, the probability $f(p)$ of the pattern p is given by $f(p) = \sum_{j=1}^{c} \alpha_j \phi(p \mid j)$. Moreover, by Bayes' theorem (also in Chapter 1), it follows that the probability an individual with pattern p on the manifest variables belongs to class j of the latent variable z is $\psi(j \mid p) = \alpha_j \phi(p \mid j) \div f(p)$.

Thus, if we can estimate the parameters α_j and θ_{ij} for all i and j then we can estimate $\psi(j \mid p)$ for each individual, assign each individual to the class for which the probability is highest and then attempt to interpret these classes in relation to the particular study. But, of course, since we have formulated probabilities of obtaining each pattern p then it is a simple matter to obtain the likelihood of the data (given by the product of $f(p)$ terms for the patterns of all individuals in the data). The maximum likelihood estimates of the parameters are then the values that make the likelihood as large as possible. This maximisation process is again an iterative numerical one (in the same spirit as the parameter estimation in factor analysis), but presents no problems with modern computer software.

Extensions of the above latent class model to the case of multicategorical manifest variables presents no further conceptual problems but simply involves more parameters. If manifest variable x_i has d categories instead of just two, then we need $d - 1$ probabilities θ_{isj} for $s = 1, \ldots, d - 1$ of the categories and probability $1 - \sum_{s=1}^{d-1}\theta_{isj}$ for the last category. Otherwise, everything is the same as before.

The same general idea can be applied if the manifest variables are quantitative rather than categorical, but now there are two sources of complication. One is that discrete probabilities such as θ_{ij} have to be replaced by probability density functions, with all the extra mathematical complications of differential and integral calculus that this brings with it. The other is that instead of just one possible formulation of the probabilities as above we have many, depending on which probability model we assume for the distributions of the manifest variables. So we do not pursue these ideas any further here, beyond saying that such models are generally known as *latent profile models*, and that if normal distributions are assumed for the manifest variables then the method is essentially the same as normal mixture separation in cluster analysis, which is discussed further in the next chapter.

9 Exploring Complex Data Sets

Introduction

Complex data sets can arise in a variety of ways. One is when the sample size is relatively small but many interrelated variables are measured on each individual, so that the associations between the variables obscure any patterns that might exist among the individuals. In other situations, fewer variables may be present but sample sizes may be too big for easy assimilation. In yet other cases, of course, we may have very many variables as well as large samples. The collection of large data sets is now commonplace, partly because of the development of increasingly sophisticated recording devices and partly because it is possible to store massive data sets electronically in such a way that instant retrieval and sorting is no problem. Bulging filing cabinets and trays of punched cards, with all the associated delays and problems involved in extracting information, are now a thing of the past. However, while the instant access to unlimited data is an undoubted benefit it also provides many problems, not least when it comes to sifting through the data in the hope of detecting interesting features or patterns within it, or of generalising from portions of it to the wider context. Of course such problems of detecting patterns are not limited exclusively to large samples; they arise surprisingly often even with small to medium sized ones.

A typical set of such data involves the observations made on a collection of variables $x_1, x_2, \ldots, x_k$ say, on each of a number n of individuals. These $n \times k$ values are most usefully arranged as a rectangular block having n rows and k columns, known as the *data matrix*. Each row corresponds to an individual and each column to a variable, so that the value in the ith row and jth column, traditionally denoted as x_{ij}, is the observation made on x_j for the ith individual. The WPPSI data discussed in Chapter 8 is a (small) example, in which there are 75 rows and 10 columns. Modern data bases might contain tens of thousands of rows and hundreds of columns, so would be stored in a computer memory and never be physically written down in such rectangular fashion, but nevertheless it will help in what follows to think of the data as represented in this way.

A researcher might have a number of objectives in analysing such a set of data. In the absence of any prior knowledge about its structure or the populations from which it has come, these objectives are likely to be purely exploratory. Are there any interesting patterns present, such as relationships between variables or distinct groupings of individuals? Are there any odd-looking or anomalous observations? Can we partition the whole set into subgroups of 'similar' individuals? These can all be thought of as *hypothesis-generating* questions, for which descriptive techniques of analysis are needed and answers to which will provide the basis of further investigation. On the other hand, there may be some *a-priori* structure imposed on the data, such as predefined groupings of either individuals or variables, and the researcher may be interested in more focused objectives. Are the groups of individuals significantly different from each other? Can a predictive rule be found that will allocate future individuals to their correct groups? Are the different groups of variables significantly associated? These are all more formal *hypothesis-testing* questions, for which inferential techniques are required.

It is evident that for even a small data set such as that of the WPPSI test scores, containing 750 numbers, simple inspection of the numbers will not be adequate. It is difficult enough to compare one complete row of numbers with another, let alone to do this for a whole set of rows. Moreover, the associations that exist between every pair of variables provide an additional obstacle and can obscure patterns to the naked eye. For example, there might be an individual in the set whose observed values all look perfectly consistent with the rest of the data values but run counter to the pattern of associations between the variables. Such an anomalous observation would escape detection in the absence of more formal analysis.

So we need a battery of analytical tools to dissect multivariate data sets, and the objective of this chapter is to provide the reasoning behind the most common methods employed in practice. However, since many of the techniques rely heavily on the mathematics of linear algebra, we will not go into any details but will restrict ourselves to a more general overview. For this reason, and for reasons of space given the many available techniques, illustrative examples will perforce be brief and superficial. We leave the reader to consult a specialised text for more detailed discussions and examples (see the list of further reading at the end of the book), and concentrate mainly on outlining the rationale of the techniques. Nevertheless, the hope is that the reader will gain sufficient appreciation of what is involved in order to be able to use some of the techniques (available in all standard software packages) with a degree of confidence. The observations need to be quantitative for most of the techniques that we discuss, so we assume that they are so unless otherwise stated.

Dimensionality and simplification

If k variables $x_1, x_2, \ldots, x_k$ have been measured on each individual in a multivariate data set, then there are k separate pieces of information about each individual, so we say that the data are k-*dimensional*. However, if all pairs of variables are correlated then there is some 'overlap' between these k pieces of information, so the 'effective' dimensionality of the data is actually less than k. As the correlation between pairs of variables increases so the overlap increases and the effective dimensionality decreases. In the extreme (and unlikely) situation where all pairs of variables are perfectly correlated (e.g. if x_1 measures the annual salary of employees in a company, x_2 measures their monthly salary and x_3 measures their weekly salary), then there is only one essential piece of information common to all variables and the effective dimensionality is 1. In practical situations, the true dimensionality will lie somewhere between 1 and k—closer to 1 when all variables are moderately to highly correlated, and closer to k when the correlations are all weak.

Dimensionality is the complicating feature of multivariate data, so wherever possible we want to simplify analysis by *reducing* the dimensionality. We will usually have some general objective in mind for the analysis (e.g. to see if there are any groups of individuals present in the data; to show up differences between predefined groups as well as possible; or to identify any outliers in the data). To achieve this objective, we want to use all the information in the original variables, but if possible to have many *fewer* variables than k to analyse and interpret. Moreover, interpretation will be aided considerably if these variables are mutually uncorrelated, because then we can consider each one separately without worrying about potential overlap in the information that each carries. Consequently, many multivariate techniques look for sets of mutually uncorrelated variables that are formed in a simple manner from the original ones and that highlight the specific objective that is of primary interest.

The simplest, and hence most common, way of condensing the information in a set of variables $x_1, x_2, \ldots, x_k$ down to a single number u, say, is to form a *linear combination* of them: $u = a_1 x_1 + a_2 x_2 + \cdots + a_k x_k$, where the a_i are all given constants. For example, the sum of a set of numbers is just such a linear combination of them with all a_i equal to 1, while their mean is a similar linear combination but with all a_i equal to $\frac{1}{k}$. So if we apply this linear combination to each row of the data matrix then we will summarise each individual in the sample by a single value—its value on the new variable u. However, it would be too much to hope that this single variable captured all the information content in the original set of variables $x_1, x_2, \ldots, x_k$. The discussion above suggests that this would only be the case if all these original variables were perfectly correlated. So in practical applications, there will be information 'left over' after we have obtained the new variable u. In order to 'mop up' all the available information, we will therefore need to find some more linear combinations:

$$v = b_1x_1 + b_2x_2 + \cdots + b_kx_k \quad \text{for constants } b_i,$$
$$w = c_1x_1 + c_2x_2 + \cdots + c_kx_k \quad \text{for constants } c_i,$$

and so on, until we can validly assume that any remaining information is equivalent to 'noise' and can be ignored.

The 'best' choice of values of the coefficients a_i, b_i, c_i, etc. will depend of course on the data but also on the particular objective that is the researcher's primary interest.

This can be accommodated by defining a suitable objective function, which quantifies the degree to which the desired objective is satisfied, and finding the values of the coefficients that optimise this objective function. We will discuss some possible functions below. However, we still have to deal with the question of correlations between the pairs of the new variables. Sometimes the requirement of uncorrelated new variables can be built into the objective function and does not need separate consideration. In other cases it can be achieved by requiring the linear combinations to be *orthogonal*. We have already met this term in Chapter 6 in connection with regression and analysis of variance. Two linear combinations

$$u = a_1x_1 + a_2x_2 + \cdots + a_kx_k \quad \text{and} \quad v = b_1x_1 + b_2x_2 + \cdots + b_kx_k$$

are orthogonal if the sum of the products of their corresponding coefficients is zero, that is, $a_1b_1 + a_2b_2 + \cdots + a_kb_k = 0$. In fact, the genesis of orthogonality in regression and analysis of variance is implicitly the same, but the underlying mathematics was not emphasised earlier. The essential point is that orthogonality of linear combinations ensures that they are uncorrelated.

We consider below some of the more common objective functions that we might wish to optimise with respect to linear combination coefficients, and the features of the data that they highlight. However, first we digress briefly regarding visualisation of multivariate data. We have seen the utility of scatter diagrams in Chapter 6 as a prelude to simple linear regression. If there are just two variables x_1 and x_2 in our data set, then we can plot a scatter diagram with these two variables as axes and this gives us a direct visualisation of the data in which the points represent the individuals. This scatter diagram shows us exactly what is going on in the data; for example, the scatter diagram in Figure 9.1 for some hypothetical bivariate data with variables x and y immediately highlights the presence of three distinct groups of individuals, in one of which there is a near-linear relationship between x and y, and one anomalous observation far away from the rest.

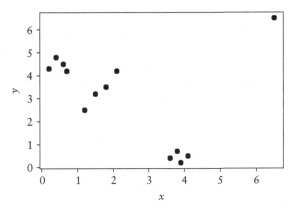

Figure 9.1.   Scatter diagram of some hypothetical bivariate data.

If we have many variables, then we might imagine each variable to be associated with an axis in space, and a 'generalised' scatter diagram of points representing individuals to be plotted against these axes. With k variables there are k axes and hence a k-dimensional plot. Of course we can never actually *see* these points, as we are limited to seeing at most in three dimensions. However, we might look for a small-dimensional space (e.g. two-dimensional or three-dimensional) that is a good approximation in the sense that when the points are projected (i.e 'squeezed') into this space the overall configuration retains its essential features and is not distorted very much. Linear combinations of the variables correspond to lines, or directions, in this space, and orthogonal linear combinations correspond to lines at right angles to each other. So a set of orthogonal linear combinations defines such a small-dimensional space into which the points might be squeezed. In particular, a pair of orthogonal linear combinations defines a two-dimensional space, and the scatter diagram obtained by plotting the points relative to such a pair of linear combinations as axes provides a two-dimensional approximation to the configuration of points in k dimensions. Such a visualisation is central to some of the following techniques.

No *a-priori* structure: principal component analysis

By far the most common need for descriptive multivariate analysis arises when the investigator has gathered a multivariate data set but has no preconceived ideas about its structure and just wishes to be able to examine the data in the best way possible in order to detect any patterns that may be present, to identify the main sources of variability in the data and to determine its effective dimensionality. Let us consider each of these objectives in turn.

The question of determining the best possible way of looking at the data reduces to the question of what is the objective function that will determine the best projection of the configuration of points from the notional k dimensions as described above, to a manageably small number of dimensions r (typically two or three) that can be plotted and inspected. Since the points have to be projected (i.e. squeezed) into a lower dimensionality, they all have to be moved somewhat from their original positions. If the configuration is not to be distorted, they should not be moved much. So the best projection is the one in which they are moved the *least*. With adherence to the well-tried principle of least squares, therefore, the best projection will be the one that minimises the sum of squared perpendicular projections of the points into the smaller space. Invoking Pythagoras' theorem as generalised to more than two dimensions (see the section below on multidimensional scaling), it turns out that this projection is equivalent to the one that maximises the spread of points in the smaller space. Since 'spread' is equivalent to 'variance' in statistical terminology, we therefore need to find the coefficients of the orthogonal linear combinations that successively maximise the variance of the points along them. Principal component analysis is the technique that achieves this, and the required linear combinations are known as the principal components.

A typical principal component routine in a standard software package will provide the user with: the coefficients of each principal component (up to a maximum of k components); the values of each individual on each component, known as the *scores* on each component; and the variances of these scores for each component. The output is arranged so that the component with largest variance comes first, the one with next largest variance comes second, and so on. Since the computer algorithm depends on a mathematical technique known as eigen decomposition, the component coefficients are sometimes referred to as 'eigenvectors' or 'latent vectors', and the variances are correspondingly called 'eigenvalues' or 'latent roots'. Also, 'loadings' is sometimes used as a synonym for 'coefficients'. The best two-dimensional configuration for viewing the data is given by the first two principal components, so the scatter diagram obtained by plotting the scores on the first two components against each other gives the best picture to examine for patterns. However, the scores on the third, fourth and succeeding components may also need to be considered if the effective dimensionality is greater than 2 (see below).

The second objective above, that of determining the main sources of variation in the data, is provided directly by the analysis as the components are arranged in decreasing order of their variances. Thus the most important source of variation is given by the first component, the next most important by the second, and so on. However, to identify these sources of variation we must interpret each component in terms of the substantive application—a process known as *reification*. This process is very similar to the one of factor interpretation in factor analysis, as illustrated by the example on WPPSI test scores in Chapter 8. We ignore 'small' coefficients on each component, and try to interpret the component in terms of those variables corresponding to the 'large' coefficients. The one extra complication in principal component analysis is that the requirement of orthogonality of components means that all except one component will include a mixture of positive and negative coefficients. Such linear combinations are known as *contrasts*, and are often harder to interpret than straightforward combinations whose coefficients all have the same sign. The way forward is to imagine what type of individuals would feature at the extremes of such contrasts. For example, if in an analysis of various body measurements of adult males, a component had a high positive coefficient for 'chest measurement', a high negative coefficient for 'waist measurement' and negligible coefficients for all other variables, then at the positive end of this component we would have males with large chests and small waists, while at the negative end we would have males with small chests and large waists. Clearly this component measures 'body shape' in general terms, or perhaps more specifically something like 'athleticism'. While such an example may afford easy interpretation, many components in practice are rather more complicated and facility with interpreting components definitely comes with experience.

The question of what is the effective dimensionality is generally tackled by looking at the variances associated with each component. The sum of the variances for all components (often referred to in software as the 'trace') gives the total variance of the data. So the ratio of the variance of each component to this total gives the proportion of overall variance accounted for by that component, and the cumulative sums of this proportion give the proportions accounted for by the first, the first two, the first three components, and so on. A common way of judging the effective dimensionality (but by no means the only way) is to say that it is the number of components required to pass a suitably large proportion—typically around 0.75–0.8. This means that around 80% of the variability of all the data is contained within this dimensionality, and the remaining 20% is small enough to be ignored. In many practical applications such a proportion is reached at fairly low dimensionalities, so considerable simplification of data structure can often be achieved.

However, there is one cautionary note that has to be sounded. If the original variables measure very different entities (e.g. one measures weights in grams, another one heights in centimetres, a third is a count of some sort, and so on) then it may not make substantive sense to form a linear combination of them. Likewise, if one or two variables have far larger variance than all the others, then they will dominate the early principal components and the analysis will not add much to existing knowledge about the data. In such circumstances it makes sense to standardise the data before the analysis is conducted, in order to put all the variables 'on an equal footing', and this means using the 'correlation' rather than the 'covariance' option on standard statistical packages. The problem is that the two analyses give very different results, a situation already encountered in the least squares factor analysis of Chapter 8 (which uses the principal component algorithm as part of its iterative process). Unfortunately, there is no equivalent of a maximum likelihood approach that gives just a unique answer, so in principal component analysis it is necessary to make a careful choice at the outset as to whether the data should be standardised or not.

To illustrate the various ideas, consider some data collected on third-year undergraduate statistics students over a three-year period (data kindly supplied by Professor Byron Morgan). There were thus 3 cohorts of students, with 51 males and 36 females in total, and each student provided measurements of their chest (x_1), waist (x_2), hand (x_3), head (x_4), height (x_5), forearm (x_6) and wrist (x_7). Although the measurements were all in the same units (inches), the ranges are somewhat different and this might be a reason for standardising the variables before analysis. However, if we simply want to inspect and interpret the data as they are, then analysis of the raw numbers is perfectly acceptable.

Principal component analysis showed that the first component accounted for 83.7% of the variance and the second for a further 7.1%, so a two-dimensional representation should be a good approximation to the original seven-dimensional configuration. The first two principal components were the following linear combinations:

$$u_1 = 0.44x_1 + 0.58x_2 + 0.14x_3 + 0.11x_4 + 0.64x_5 + 0.15x_6 + 0.10x_7,$$

$$u_2 = -0.39x_1 - 0.56x_2 + 0.08x_3 + 0.09x_4 + 0.71x_5 + 0.12x_6 - 0.01x_7.$$

Thus u_1 is a weighted sum of the separate variables, with major emphasis on chest, waist and height measurements. Since 'large' students will have large values of x_1, x_2 and x_5, and hence of u_1, while 'small' students will have small values, then u_1 can clearly be interpreted as a measure of 'body size'. On the other hand, while u_2 also focuses on the same three variables, the chest and waist measurements are now subtracted from the height measurement. 'Large' values of u_2 correspond to students whose heights are relatively greater than their chests and waists, while 'small' values will correspond to students who have the reverse characteristics. So u_2 is a measure of 'body shape'. But notice that this is very much the secondary characteristic that distinguishes the students (7.1% variance for u_2 as against 83.7% for u_1).

To examine the data for any interesting features, we can therefore plot the scatter diagram of scores on u_1 against those on u_2. This plot is shown in Figure 9.2, and each point is labelled according to the gender (upper case for male, lower case for female) and cohort (a, b, c for first, second and third) of the relevant student. Immediate deductions that follow from the plot are that the first principal component distinguishes the males from the females very effectively (with only a small amount of intermingling at

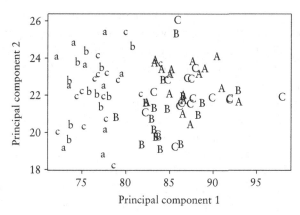

Figure 9.2. Scatter diagram of the student data on the first two principal components.

the border), that there is considerable intermingling of the three cohorts and that there is one clear outlier on the first principal component (a very large male!)

Grouping of the individuals: canonical variate analysis

Frequently in practice there is some *a-priori* structure on the data, which must be taken into account in the analysis. The most common such structure is when the individuals fall into a number of predetermined groups, and the purpose of the analysis is to highlight any differences that exist between these groups. For example, suppose that a psychiatrist has classified each of the n subjects taking part in a study of mental health either as 'well' or as suffering from one of a number g of psychiatric disorders. Thus there are $g + 1$ *a-priori* groups into which the subjects have been placed. Each subject is then asked to give a score on a set of questions, such as: x_1—have you recently felt that you are playing a useful part in things? x_2 —have you recently felt contented with your lot? x_3—have you recently felt capable of making decisions about things? x_4—have you recently felt you are not able to make a start on anything? The researcher might then be interested in determining (i) whether the groups differed with respect to these questions, (ii) which combinations of questions best separate the groups, and (iii) whether any groups identify with specific questions.

Such enquiries can be answered with the help of a descriptive technique that enables us to plot the data in a low dimensionality, in such a way as to highlight group differences. Principal component analysis will not usually be appropriate, however, as it is concerned simply with overall variance of the individuals and pays no regard to the group structure, so is quite likely to produce a representation that obscures rather than highlights the group differences. Instead, we need to generate a different representation, by including the difference between groups in the objective function that is to be optimised.

The first attempt at tackling such a problem was made by Sir Ronald Fisher in the 1930s, for the special case where there are just two *a-priori* groups. As with much of Fisher's innovative work he based his approach on intuition, and also typically produced a solution which was not only subsequently shown to be optimal under various formal statistical assumptions, but which also opened up the way for many generalisations. Fisher's idea was very simple: to look for the linear combination $u = a_1x_1 + a_2x_2 + \cdots + a_kx_k$ of the original variables such that when the groups are represented on this new variable u they are 'as distinguishable as possible'. If we know the values of the coefficients a_i then we can find the value of u for each individual in the sample, in which case the original multivariate data set has been converted into a univariate one in which the individuals are split into two groups. The significance of the difference between the means of two groups can be tested by the two-sample t-statistic, which is the ratio of the difference between the two sample means to the standard error of this difference. Calculation of this standard error usually assumes a common variance in the populations from which the groups have come, and its value is a simple multiple of the pooled within-group standard deviation. So the square of the t-statistic is proportional to the squared difference between the two sample means divided by the pooled within-group variance. The bigger the value of this statistic the more separated are the groups, so this was the objective function that Fisher set out to optimise.

His solution, now universally known as *Fisher's linear discriminant function (LDF)*, is thus the linear combination $u = a_1x_1 + a_2x_2 + \cdots + a_kx_k$ whose coefficients are the ones that maximise the criterion $W = (\bar{u}_1 - \bar{u}_2)^2/s_u^2$, where $\bar{u}_i$ denotes the mean of the u values (i.e. scores) for individuals in group i and s_u^2 is the pooled within-group variance of these scores. There is a fairly straightforward formula from which the coefficient values can be calculated, utilising the group mean vectors and the pooled within-group covariance matrix. However, Fisher also showed that they can be obtained by a multiple regression of y on the original set of variables $x_1, x_2, \ldots, x_k$, if y is a group indicator variable having one value for each individual in the first group and a different value for each individual in the other group. Any two distinct values can be used for y. The simplest choice is 0 and 1, but another common one is $\frac{n_2}{n_1 + n_2}$ for the n_1 individuals in group 1 and $-\frac{n_1}{n_1 + n_2}$ for the n_2 individuals in group 2. The latter choice is convenient for predictive purposes, as it is the one for which the mean of all the u values is zero.

The extension of Fisher's idea to the general case of g groups came in the 1950s, and while at the time it was innovative we might now be tempted to consider it as fairly obvious. As before, if we know the values of the coefficients a_i then we can calculate u for each individual in the sample and can thereby convert the original multivariate data set into a univariate one. However, now the individuals are split into g groups, and the significance of any differences among the means of g groups is tested by a one-way analysis of variance. Recollect from Chapter 6 that the requisite test statistic here is the F ratio, which is the between-group mean square divided by the within-group mean square. By varying the values of a_i we obtain different sets of u values and hence different values of this F ratio. Clearly, the best combination for showing up group differences is given by the one whose coefficients *maximise* this ratio.

However, whereas in the case of two groups there was just one linear combination to show up the differences (so the problem was one-dimensional), in general with g groups the problem is $(g-1)$-dimensional and there will be $(g-1)$ combinations. This is similar to the principal components situation, so to distinguish the techniques we call these linear combinations the *canonical variates*. The first canonical variate is thus the linear combination of the original variates $u = a_1x_1 + a_2x_2 + \cdots + a_kx_k$ that maximises the F ratio of between-group mean square to within-group mean square. The second canonical variate is the linear combination $v = b_1x_1 + b_2x_2 + \cdots + b_kx_k$ that is uncorrelated with the first and has the next largest F ratio, and so on for further canonical variates with each new one being uncorrelated with all preceding ones (but it should perhaps be noted that although the canonical variates are mutually uncorrelated they are not mutually orthogonal). As in principal component analysis, computer packages will provide several quantities: successive values of the F ratio (sometimes labelled 'canonical roots' or, as before, 'eigenvalues'); the sum of all F values ('trace') plus percentages of this sum for each F; coefficients of the canonical variates (sometimes labelled 'eigenvectors' or 'latent vectors'); and scores on each canonical variate for each sample member.

We would again look for sufficient number of canonical variates in order to account for 80% or more of the trace, and plot the data against these canonical variates as axes in order to view the individuals in the way that best shows up differences between groups. For this purpose it is useful to plot the points corresponding to individuals in different groups with different symbols. Another very useful display is just to plot the group means against the canonical variates as axes. Such a plot removes the clutter caused by a large number of individual points, and focuses attention on the relationships among the group means. In addition, there are simple formulae that enable either confidence regions or tolerance regions to be constructed around each group mean point, assuming multivariate normality of the data. The former indicate regions within which we expect the population means to lie with given degree of confidence, while the latter indicate regions within which given proportions of the whole population are expected to lie.

As an illustration, let us return to the student data previously subjected to principal component analysis. That analysis concentrated on differences between the students and paid no heed to any groups that might be present, but in fact there are actually six distinct groups (males and females, each divided into three cohorts). The principal component analysis managed (as a by-product) to show up differences between the genders, but not between the cohorts. We might therefore hope to do better by specifying the group structure and applying canonical variate analysis. When we do this we find that the first canonical variate is

$$v_1 = 0.00x_1 + 0.32x_2 + 1.39x_3 + 0.08x_4 - 0.12x_5 + 0.14x_6 - 0.19x_7$$

and it accounts for 64.3% of the trace, while the second canonical variate is

$$v_2 = -0.16x_1 + 0.34x_2 - 1.47x_3 + 0.15x_4 + 0.24x_5 - 0.11x_6 + 0.95x_7$$

and it contributes a further 33.1% of the trace. Thus a two-dimensional scatter diagram of v_1 against v_2 should give a very good approximation of the representation that shows up differences between the six groups, and this scatter diagram is given in Figure 9.3.

We see that the orientation is a little different from the one in Figure 9.2, the boundary between males and females being a sloping line roughly joining the points marked 20 on the two axes rather than the previous vertical line in the middle of the horizontal axis. Such a change of orientation is to be expected, as we are providing an essentially different view of the data from that of the principal components. However, while there is just as clear a separation of males and females as before, we now also have a much more evident division of each gender into its three cohorts. The third cohort of each (groups 'c' and 'C') is markedly separate from the other two, and while

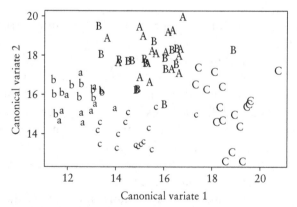

Figure 9.3.   Scatter diagram of the student data on the first two canonical variates.

there is rather more overlap between the other two cohorts for both genders these groups are more clearly distinguished than in Figure 9.2.

Grouping of the variables: canonical correlation analysis

Sometimes it is the *variables* in a study instead of the individuals that are divided into *a-priori* distinguishable groups, and the investigator is interested in determining the associations that exist between the *groups* rather than between individual pairs of variables. For example, in a battery of IQ tests, some might be 'verbal' in nature (e.g. finding words to complete sentences, identifying the word in a group of words that is the 'odd one out', and so on) while others are 'numerical' (e.g. specifying the next number in a sequence, doing arithmetical tasks, and so on). The educational researcher is less interested in specific tests, but more broadly wishes to investigate connections between 'verbal' and 'numerical' abilities. In a study of pilots' reactions to high altitude, subjects are observed doing some task in a decompression chamber under different pressures. For each subject a number of physiological responses are measured (e.g. blood pressure, heart rate, finger tremor, and so on) and a score is elicited on various psychological questions (e.g. how uncomfortable is it, do you feel agitated, do you think you are succeeding in the task, and so on). Here again, the investigator is less interested in specific variables than in the broad connections between how the subjects are reacting (the physiological variables) and how they *think* they are reacting (the psychological ones).

The problem with data from such observations or experiments is that the correlation matrix is the natural focus of attention, but this matrix can give a very confusing picture. This is partly because there are very many values in it, but more pertinently because it contains many 'within-group' as well as 'between-group' correlations. So it is not easy to discern what the broad picture is—we cannot see the wood for the trees. A technique that attempts to 'clean up' the correlation matrix in such a way that the main connections between groups of variables can be seen is *canonical correlation analysis*.

For a brief summary of this technique, let us suppose that the variables in the first group are denoted $x_1, x_2, \ldots, x_p$, while the variables in the second group are denoted $y_1, y_2, \ldots, y_q$. Canonical correlation analysis (also available on all standard computer packages) produces *two sets* of linear combinations, one set for each group of original variables: $u_1, u_2, \ldots, u_s$ are all linear combinations of $x_1, x_2, \ldots, x_p$, while $v_1, v_2, \ldots, v_s$ are all linear combinations of $y_1, y_2, \ldots, y_q$, where s is the smaller of p and q. Moreover, the u_i and v_j have the following properties:

(1) all the u_i are mutually uncorrelated;
(2) all the v_j are mutually uncorrelated;

(3) every u_i is uncorrelated with every v_j apart from its partner v_i;

(4) the largest correlation r_1 is between the first pair u_1, v_1, the next largest r_2 is between the second pair u_2, v_2, and so on.

Thus, if the u_i and v_j are considered as new variables then the correlation matrix for them takes a particularly simple form. It has 1s down its principal diagonal (of course), but all other entries are zero except those between corresponding u_i, v_i pairs where the entry is r_i. These r_i are the maximum possible correlations between linear combinations of the two sets of original variables, subject to the constraints in (1)–(3) above, so they quantify the 'pure association' between the two groups of variables. In other words, they give the correlations 'between' groups after removing the effects of the correlations 'within' groups. It is possible to test whether each r_i is significantly different from zero, assuming a multivariate normal distribution for the data, and if a computer package has a canonical correlation option then it will deliver the outcome of this test. The number of significant such correlations is thus the dimensionality of association between the two groups.

The r_i are called the *canonical correlations* of the system, while the u_i and v_j are also known as *canonical variates*. This is because if $y_1, y_2, \ldots, y_q$ is a set of dummy variables giving the group membership of each individual in a data set having grouping of individuals (see Chapter 6), while the $x_1, x_2, \ldots, x_p$ are a set of measured variables, then a canonical correlation analysis will produce the same results as a canonical variate analysis of the $x_1, x_2, \ldots, x_p$.

Scaling methods

In all of the above techniques we have assumed that the variables $x_1, x_2, \ldots, x_k$ are quantitative, and this has allowed us to connect the n individuals in the data set with n points in k-dimensional space by setting the coordinates of the ith point equal to the values on each variable for the ith sample individual. Implicit in this geometrical model is the idea that distance between points indicates dissimilarity between corresponding individuals: Two points close together indicate two individuals having *similar* values on each variable, while two points far apart indicate two individuals that are very *dissimilar* with regard to the measured variables.

Measurement of ordinary (i.e. Euclidean) distance between points rests on Pythagoras' theorem. Consider the two-dimensional space in Figure 9.4, with points A and C having coordinates (a, b) and (c, d) on axes x and y, respectively. The distance D, say, between these points can be obtained easily by constructing the right-angled triangle shown, with the addition of point B whose x coordinate c is the same as that of point C and whose y

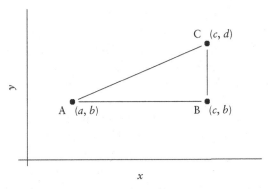

Figure 9.4. Distance between two points in a plane.

coordinate b is the same as that of point A. Then by Pythagoras' theorem D^2 is the sum of squared lengths of AB and BC. But since AB is parallel to the x-axis its length is just $(c - a)$, and likewise with BC parallel to the y-axis its length is just $(d - b)$. So $D^2 = (c - a)^2 + (d - b)^2$, and D is the square root of this. When we pass to k-dimensional space we use the obvious generalisation, so that if one point has coordinates $(x_1, x_2, \ldots, x_k)$ and another point has coordinates $(y_1, y_2, \ldots, y_k)$ in this space, then the squared distance between them is given by

$$D^2 = (y_1 - x_1)^2 + (y_2 - x_2)^2 + \cdots + (y_k - x_k)^2.$$

Thus, whenever we have a configuration of n points in a k-dimensional space, we can use this formula on the coordinates of the points to calculate the distance between every pair of points. These distances can be placed in a matrix having n rows and n columns, such that the value d_{ij} in the ith row and jth column is the distance between the ith and jth points. Such a matrix has two distinctive features: it is *symmetric* (i.e. $d_{ij} = d_{ji}$, since the distance between the ith and jth points is clearly the same as the one between the jth and ith points), and the values d_{ii} down its diagonal are zero (as the distance between any point and itself must be zero). There is a nice mathematical result, which establishes that we can carry out this process in reverse: If we are given such a matrix of all inter-point distances, it is easy to construct from it a configuration of points that give rise to these distances. Moreover, the mathematics again involves eigen decomposition of a matrix so that this configuration is referred to its principal axes, that is, it is the configuration arising from a principal component analysis. The benefit of having it in this form is that the dimensions are generated in their 'order of importance', the first having the greatest spread of points, the second the next greatest, and so on. So if we want a low-dimensional approximation to the configuration, say in r dimensions, we simply need to take the first r axes of the recovered configuration. Because of the relationship with principal component analysis, this technique is sometimes called *principal coordinate analysis*. More usually nowadays, however, it is referred to as *metric scaling*.

The real benefit of the technique is that, if we associate dissimilarity between individuals with distance between points, and we are able to calculate a matrix of dissimilarities between every pair of individuals in a sample, then we can use the technique to produce a configuration of points in such a way that the distance between any two points approximates the dissimilarity between the corresponding individuals. This extends the possibility of looking at geometrical representations to situations where the multivariate data include nominal or categorical variables, which were not covered by the previous techniques, and enables us to seek patterns or detect outliers in such data as well. There are now many dissimilarity formulae that will enable us to calculate an appropriate dissimilarity matrix with any data set. For example, a simple dissimilarity measure between individuals on the basis of a set of categorical variables is just the proportion of variables on which the individuals exhibit *different* categories. The connections noted above with principal component analysis mean that the earlier guides to interpretation of configurations (such as methods of determination of effective dimensionality) that were appropriate in the principal component case are equally applicable here. There are some extra mathematical complications in certain situations, and there are variations not only on the way in which the configurations are derived but also on the method of approximating dissimilarities by distances.

However, going into details is beyond our brief; suffice it to say that the technique opens up descriptive analysis to virtually all types of data situation. As a brief illustration, consider a study that was made into the voting patterns of some US congressmen. Fifteen members of the US House of Representatives from New Jersey were monitored, and their voting patterns on 19 environmental bills were recorded. A very simple measure of dissimilarity between two congressmen was defined to be the number of times that they voted differently on these bills, and this enabled a 15×15 matrix of dissimilarities between every pair of congressmen to be drawn up and subjected to metric scaling. About 70% of the differences between congressmen was represented on the first two scaling axes, so this should provide a reasonable approximation to these differences. The resulting scatter diagram is shown in Figure 9.5, the points being labelled according to the party allegiance of the congressmen (D = Democrat, R = Republican).

Some interesting patterns are immediately apparent from this diagram. There is a large difference between the two party groups, indicating that environmental issues seem to be a major source of differences between the parties. Broadly speaking, the Democrats bunch more closely together than the Republicans, perhaps indicating a more stringent 'party line' on such issues, although both parties also have one or two 'outliers'. One of these outliers (the Republican in the midst of Democrats) seems to be so extreme as to

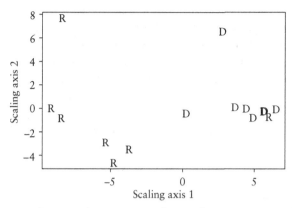

Figure 9.5.   Scatter diagram of congressmen data on first two metric scaling axes.

suggest a mistake in either the data collection or the labelling of that individual. The other outliers are perhaps explained by the fact that a few congressmen abstained from voting on a number of occasions. It is debatable whether these observations would have come to light by simple inspection of the raw data.

Unsupervised classification; cluster analysis

Constructing configurations of points representing individuals may be a good way of looking at a dissimilarity matrix in order to investigate patterns, but it suffers from a number of potential drawbacks in any given application. The configuration necessary for exact representation may require many dimensions, so that any approximation for visualisation in two or three dimensions may be very poor and may thus obscure rather than reveal pattern. Moreover, if the objective is to identify groups of 'similar' individuals, the recovered configuration may not reveal them very clearly. Even if a large proportion of variability is recovered in the first few dimensions, the crucial divisions between groups may be present in one of the higher dimensions and may thus be missed.

A desire to group individuals into more or less homogeneous classes has been central to much human activity over the years (even centuries!), and such *classification* is of prime importance in many fields. One purpose is to bring a sense of order to large collections of disparate entities, in order to aid identification and recognition. For example, biologists have classified organisms, plants or animals using labels such as genus, species, class, and so on for many years, so that they can be easily identified, distinguished and categorised. Library books are classified according to their content using a system like the Dewey decimal system, in order to make it easier for a user to find the books he or she wants. Another purpose may be to simplify data structure, or to break a set of individuals down into subgroups in order to optimise further action. For example, market researchers might seek classes into which they can place consumers according to their product preferences, so that they can target new products at an appropriate retail market. Indeed, in virtually any quantitative field at the present time, the objective of classifying multivariate observations into groups is an important one.

We may note in passing that there are two distinct types of situations that can be envisaged: situations in which distinct and relatively homogeneous groups are believed to exist, and the objective is to find them; and situations in which we simply wish to divide a possibly large data set into a number of groups for convenience, where individuals within any group are fairly similar while individuals between two groups are rather different. The former type of situation is the one usually found in Biology, where the different classes such as species or genera are intrinsically distinct; while the latter type is the one usually found in market research, where the classification is for convenience of retailing rather than there being any intrinsic groups that need to be found. However, while there are two underlying types of situation, they generally only provide a reason for the study rather than guiding the methodology. With one or two exceptions, any of the existing methods can be applied to either type.

Numerical approaches to the grouping of individuals date back at least to the early nineteenth century, and were systematised in the work of numerical taxonomists concerned with the classification of organisms. While 'classification' was a suitable description of this activity, the word came to have two distinct scientific connotations: not just the division of individuals into (previously undefined) classes, but also the placing of (unidentified) individuals into pre-existing classes (see below). Thus, in order to distinguish these two connotations, statisticians came to refer to the former as *clustering* or *cluster analysis*, while in more recent times computer scientists have coined the term *unsupervised classification* to denote that there are no pre-existing classes to guide the analysis. Unsurprisingly, given the long history of development and the large number of different starting points of development, there is now a plethora of different methods of cluster analysis, not to mention computer algorithms and implementations. This is no place to go into any of the details of an enormous topic, so we shall simply look at the three major ways of tackling a cluster analysis and outline the general strategy for each approach.

The oldest of the approaches reflects its biological and taxonomic roots, in which the classifications sought were always nested ones: individuals within species, species within genera, genera within classes, and so on. This approach is known as a *hierarchic cluster analysis*. The idea here is that if one has a set of *n* individuals to be clustered, one systematically generates all the possible number of clusters from *n* (with each individual in its own cluster) through to just one (with all individuals in a single cluster), but subject to a constraint. If at any stage there are g clusters, then at the next stage either $g + 1$ clusters will be obtained by *splitting* one of the existing ones, or $g - 1$ clusters will be obtained by *merging* two of the existing ones. Thus there are two possible strategies in order to generate all the clusters: We can either start with all individuals in a single cluster ($g = 1$) and then proceed by successively splitting one cluster into two at each step until all the individuals are located in separate clusters, or we can start with all individuals in their own clusters ($g = n$) and proceed by successively merging two clusters at each step until we have a single cluster containing all individuals. The former strategy is known as *divisive clustering*, but it is very demanding of computer time so has not been popular in the past. By contrast, the latter strategy is known as *agglomerative clustering* and is generally quick and easy computationally so has been much used in practice.

The main question to settle with agglomerative clustering is which clusters to merge at each stage, and the idea of distance or dissimilarity again plays a vital role here. We start by finding the dissimilarity matrix between all pairs of individuals in the sample, so the dissimilarity matrix has n rows and n columns, and each individual is in its own cluster. The computational procedure is then as follows:

1. Find the two most similar clusters (the ones corresponding to the smallest dissimilarity) and merge them into a new cluster.
2. Recompute dissimilarities between the new cluster and each of the other clusters, thereby reducing the dissimilarity matrix by one row and one column.
3. Cycle through steps 1 and 2 until only one cluster remains.

To carry out this process we need to know how dissimilarity is calculated between two clusters in order to complete step 2. There are many ways in which this can be done, each corresponding to a different 'clustering method'. The most common ways are naturally the simplest ones that are easiest to calculate. Suppose there are a individuals in one cluster and b individuals in the other, then the dissimilarity between the two can be measured by

(1) the smallest of the $a \times b$ inter-object dissimilarities (this is the *nearest neighbour* method, also known as *single-linkage*);
(2) the largest of the $a \times b$ inter-object dissimilarities (this is the *furthest neighbour* method, also known as *complete-linkage*);
(3) the average of the $a \times b$ inter-object dissimilarities (this is the *group average* method, also known as *average-linkage*).

Of course, many other possibilities exist but the three above, between them, account for many practical applications of hierarchical cluster analysis. Once the groupings have been obtained, a complete history can be represented graphically in the so-called *dendrogram* ('tree diagram') that shows how the successive mergers have taken place.

Hierarchical clustering, and in particular the single-link method (for which there are special very quick algorithms), was the preferred method while computing resources were primitive or expensive, but has been overtaken more recently by several other approaches as computers have become both fast and cheap to run. One very popular general approach is what might be termed the 'optimisation' approach, in which a numerical criterion representing the 'goodness' of a clustering solution is defined, and then the computer is left to find the division of objects into clusters that optimises this criterion. However, in this approach the user must specify how many clusters the objects are to be divided into. So, for example, we may define the goodness of the clustering to be measured by the within-cluster total sum of squares over all variables. If the data are divided into tight and well-separated clusters then the value of this criterion will be very small, so we set the computer to find the clustering that minimises this criterion. To do this, it can try all possible division of the n objects into the specified number k of clusters, but if n is at all large this is a computation-ally infeasible task and various 'transfer' algorithms are used in order to speed the computing. These are approximate methods that move single objects between clusters in order to improve on the current value of the criterion, until no more improvements are possible. They do not necessarily find the 'best' clustering, although they usually find a 'good' clustering. The k-means algorithm is a popular 'optimisation' method of clustering, but like hierarchical methods there are many other possibilities. However, one general drawback occurs when the 'best' number of clusters has also to be determined. In this case, the optimised value of the criterion must be computed and compared for different numbers of clusters into which the objects have been placed, but there are few objective guidelines to help in this comparison and a decision must be made on mainly descriptive grounds.

A final general method worthy of mention is also one that has only become feasible since computer power became cheap and plentiful. It is what one might term a statistical, model-based approach, in which a prob-ability model that includes a group structure is postulated for the data. Parameters of the model and group memberships of the individuals are then estimated by maximum likelihood. The most common method of this type is the *normal mixture model*. Here again we have to specify the number k of clusters that are sought. Then the model assumed for the data is a mixture of k (multivariate) normal populations differing in their mean vectors but not in their covariance matrices. Each individual in the sample is assumed to belong to one of these populations, although initially the allocation of individuals to populations is unknown. However, it is possible to formulate the likelihood of the data under these circumstances, and then by using what is known as the E-M algorithm (for maximising a likelihood when there are unknown aspects in it) we can estimate the population parameters and the population membership of each individual. This is potentially a very powerful method, but carries with it both computational and statistical problems (the latter again focusing on the optimal choice of k).

Supervised classification; discriminant analysis

We now turn to the other meaning of the word classification, namely the process of allocating individuals to one of a number of predefined classes. This process is a very familiar everyday one in medicine, as no doubt most readers will have experienced visits to their doctor as a result of which they are allocated to one of a number of 'illness' or 'treatment' categories on the basis of a set of symptoms and measurements such as temperature, blood pressure, and so on. However, examples abound in many other fields: an archaeologist may have obtained measurements on skeletons found in a number of different localities and now wishes to establish the likely origins of other unidentified skeletons; a marine biologist has obtained measurements on a number of different species of phytoplankton, and now wishes to identify the species of a sample collected from a previously unexplored region of the sea; a bank has amassed information on the financial transactions of customers who have taken out loans, has categorised these customers according to their arrears in repayment and now wishes to allocate new applicants for credit into one of these categories.

There is a common pattern to all these applications: there are g, say, predetermined classes; k variables $x_1, x_2, \ldots, x_k$ have been measured on a set of labelled individuals (i.e. ones whose class memberships are known); and a *classification rule* is sought for allocating future unlabelled individuals to one of these classes on the basis of their values of $x_1, x_2, \ldots, x_k$. Since labelled individuals are available for the building of the classification rule, such classification problems may be termed *supervised* classification in order to distinguish them from the clustering problems discussed in the previous section. In common with all other multivariate techniques, the most popular line of attack is to find some optimal function or functions of the variables $x_1, x_2, \ldots, x_k$ to use as a classification rule, but how do we set about obtaining such functions? One logical way would be to focus on functions that best separate, or *discriminate between*, the predefined classes. This was the original statistical approach, so *discriminant analysis* became the accepted term and is now used as a synonym for supervised classification. The set of labelled individuals from which the rule is to be constructed is often called the *learning set* or the *training set*.

If we are looking for functions that best discriminate between groups, and we focus on the simplest type of function, namely the linear function, then we are essentially back to the descriptive technique of canonical variate analysis described earlier. We can allocate future individuals by finding their coordinates on the canonical variates obtained from the training data, plotting them on the canonical variate diagram and allocating them to the group to whose mean they are nearest. More directly, if $g = 2$ and we are allocating to one of just two classes (e.g. 'ill' or 'well' in the medical context) then we can use Fisher's linear discriminant function. If we take the multiple regression approach previously mentioned, and assign values $\frac{n_2}{(n_1 + n_2)}$ for the n_1 individuals in group 1 and $-\frac{n_1}{(n_1 + n_2)}$ for the n_2 individuals in group 2 to the dummy variable y, then a future unlabelled individual can be allocated on the basis of its predicted value $\hat{y}$ from the fitted multiple regression. Since the mean of the y values in the training data is zero with the above assignment, we allocate the individual to group 1 or group 2 according to whether its predicted $\hat{y}$ value is positive or negative.

The above approach is both simple and pragmatic, but will only be applicable to strictly quantitative data and begs the question whether it has any optimality properties for such data. A more systematic, and theoretically based, argument starts from the premise that the predefined classes are actually populations of (potential) individuals, and that the probability density function in the ith population of the variables $x_1, x_2, \ldots, x_k$ is some function $f_i(x_1, x_2, \ldots, x_k \mid \theta)$ depending on the unknown parameters θ. To take account of possibly very different sizes of populations (e.g. if we are trying to distinguish between a common respiratory illness experienced by many individuals and active tuberculosis that affects relatively few) we need to allow for differential probabilities of obtaining an individual from each population, so we let π_i be the prior probability of a randomly chosen individual being from class i (with $\pi_1 + \pi_2 + \cdots + \pi_g = 1$). Application of Bayes' theorem (Chapter 1) then tells us that the posterior probability that an individual having values $x_1, x_2, \ldots, x_k$ comes from population i is

$$p_i = \frac{\pi_i f_i(x_1, x_2, \ldots, x_k \mid \theta)}{\pi_1 f_1(x_1, x_2, \ldots, x_k \mid \theta) + \pi_2 f_2(x_1, x_2, \ldots x_k \mid \theta) + \cdots + \pi_g f_g(x_1, x_2, \ldots, x_k \mid \theta)}.$$

The optimal procedure, known as the *Bayes allocation rule*, is to classify the individual to the population whose posterior probability is highest.

Of course, these posterior probabilities are theoretical as long as the probability density functions $f_i(x_1, x_2, \ldots, x_k \mid \theta)$ and the prior probabilities π_i are unknown, so in any practical application these quantities must all be estimated. A *parametric* approach requires us to specify the functional form of the densities $f_i(x_1, x_2, \ldots, x_k \mid \theta)$, which we would do with regard to the type of variables $x_1, x_2, \ldots, x_k$ we observe, and then to eliminate the unknown parameters θ in some way. The classical (frequentist) approach is to replace them by their estimates from the training data, while the Bayesian approach is to formulate a prior distribution for them, obtain the posterior distribution from the training data, and then remove them by finding the predictive distribution (i.e. the expectation over the posterior distribution) of a new individual. The prior probabilities π_i are often known from the substantive context, but if they are not known then they can either be estimated using the observed proportions in the training data or simply all set equal if the training data were obtained by sampling a fixed number from each population. As can be appreciated from this brief description, the process can produce very many different classification rules depending on choice of models and choice of inferential approach. However, if the frequentist approach is adopted, and the populations are all modelled by multivariate normal distributions having equal dispersion matrices, then this process reduces once again to a series of linear discriminant functions. In particular, when $g = 2$, we again recover Fisher's linear discriminant function, which is another reason for its continuing popularity in practical applications.

In recent years, with access to virtually unlimited computing power, interest has switched to computationally intensive derivation of classification rules. Within the above parametric approach, Markov Chain Monte Carlo methods have enabled Bayesian classification rules to be derived for very complicated models of the *a-priori* populations. However, interest has been increasingly focused on *nonparametric* approaches in which the density functions $f_i(x_1, x_2, \ldots, x_k \mid \theta)$ are not constrained to have specific functional form but are instead estimated directly from the training data using such mechanisms as kernel density estimators, wavelet estimators or orthogonal series estimators, and on *machine learning* methods such as neural networks and support vector machines which aim to optimise the probabilities of correct classification more directly. The upshot is that the practitioner now has an enormous range of possible classification rules from which to choose.

With such an element of choice, how do we assess the performance of a classification rule so that we can choose the best one to use in any given situation? Various different criteria have been put forward over the years, but the overwhelming preference in applications has been to focus on the success rate (or its complement, the error rate) as measured by the proportion of allocations a particular rule is expected to get right (or wrong). If the rule has been derived from theoretical considerations via parametric assumptions about the probability distributions, then it is also generally possible to evaluate theoretically the probabilities of misclassification of individuals via this rule (i.e. its error rate), and then to replace any unknown parameters in these probabilities by their estimates from the training data. However, this is rarely done, because error rates derived in this way are not very robust to departures of the data from the assumed probability models, and the estimated probabilities may therefore not be very reliable. Recourse is nearly always made to data-based estimation of error rates, using the training data in some way.

The earliest approach was to simply re-use the training data, namely to form the classification rule from it and then to estimate the error rate by the proportion of the labelled individuals in the training data that were misclassified by this rule. However, it was soon realised that such a procedure of *resubstitution* gave a highly over-optimistic assessment of future performance of the rule (because the rule is perforce the one that works *best* on the training data). To get a good estimate of future performance, it is necessary to test the rule on a set of data that has *not* been involved in the rule's construction. If there is a large enough set of training data available, then part of it can be hived off and used exclusively as a *test* set for estimating the error rate. However, this is a luxury not often accorded in practice, so various ingenious methods have been devised to get round the problem of biased estimates of error rate. One is *cross-validation*, in which the training data are divided into a number h of subsets. Each one of these subsets is left out in turn, the rule is constructed from all the individuals in the other $h - 1$ subsets, and the number of individuals in the omitted subset that are misclassified by this rule is noted. When all subsets have been left out in this way the total number of misclassifications divided by the total number n of training set individuals is an estimate of the error rate of the rule formed from all training set individuals. At the extreme, for small data sets, the number of 'subsets' is set equal to n so that each individual is in its own subset, and at each iteration of the process just one individual is omitted from the data. This is often termed the *leave-one-out* estimate of error rate. Other common data-based methods are based on the idea of *bootstrapping*, where samples of size n are repeatedly drawn with replacement from the training data, and error rates can be estimated in various ways from these repeated samples.

To illustrate these ideas let us return to the student data considered earlier and see how we would fare in a prediction of group membership of the students using the measurements previously described. The method chosen was the Bayes allocation rule in its most common parametric implementation, namely for the case of normal distributions with a common dispersion matrix in all the groups, and assessment of the rule's performance was by means of leave-one-out cross-validation. When only gender was considered, so that prediction was to either 'male' or 'female', then (as might be anticipated from the earlier diagrams) the rule performed very well. Only 3 of the 51 males were classified 'female' and only two of the 36 females were classified 'male', the overall success rate thus being 94.3%. The misclassifications were nearly all the ones that obviously lay on the border between the groups in Figures 9.2 and 9.3 and were likewise near the $p_1 = p_2 = 0.5$ probability borderline in the Bayes rule, although there was one female who was very unequivocally classified as male! When it came to prediction of all six groups, then as might be expected the performance fell off somewhat. The overall success rate here was 62.1%, and the misclassification rates of individuals from each group were as follows: 7 of the 17 in A; 9 of the 17 in B; 2 of the 17 in C; 7 of the 10 in a; 5 of the 11 in b; and 3 of the 15 in c. The high proportion of successes in groups C and c is not surprising given the relative isolation and coherence of these groups as shown in Figure 9.3. Likewise, the relative intermixing of the other cohorts for each gender accounts for the poorer success rates in the other groups.

The area of supervised classification is perhaps one of the most intensely researched areas of multivariate analysis, and it is difficult to do justice to it in such a brief overview. The interested reader should follow up the recommended texts given at the end of this book.

Inferential techniques

The emphasis in this chapter has been very heavily on descriptive multivariate techniques, partly because these are the techniques most commonly encountered in applied research and partly because they are the ones that involve ideas not covered in previous chapters. This is not to detract from the large body of technical material now available for inferential applications in a multivariate setting. However, the fundamental ideas of inference are basically much the same in the multivariate setting as they were in the univariate setting, the main differences being either in the conceptual extensions necessary to carry one-dimensional ideas into higher dimensionalities, or in the mathematical technicalities necessary to manipulate vectors and matrices instead of single variables or numbers. As we are not concerned with mathematical technicalities, we will simply content ourselves here with a brief outline of the necessary conceptual extensions and the general approaches taken to achieve workable results.

All the fundamental principles and approaches discussed in Chapters 4 and 5 for univariate situations are maintained in the multivariate case also. Thus, on the frequentist side we are concerned with parameter estimation and hypothesis testing as before, and on the Bayesian side we focus on posterior distributions for inference and predictive distributions for eliminating unknown parameters. The major conceptual change is in dimensionality. Population models for a set of variables $x_1, x_2, \ldots, x_k$ are k-dimensional, so probability density functions are (notional) surfaces in the $(k+1)$th dimension rather than simple curves, and each unknown parameter in such a density function has (at least) k components. So the position of an unknown parameter is a point in k-dimensional space rather than just a value on a single axis. However, while this extension to k dimensions leads to considerable increase in the complexity of mathematical calculations, it does not affect most of the basic inferential tenets. The only conceptual change is in the extension to univariate confidence intervals for unknown parameters where, instead of an interval of real numbers within which we have a given confidence that the parameter will lie, we now have to construct a *region* in k-dimensional space within which we have a given confidence that the point representing the parameter will lie. All other basic principles remain essentially the same.

The chief task, therefore, is to find the quantities on which to base the inferential techniques, namely the estimators for parameter estimation, the pivotal quantities for confidence regions, the test statistics for hypothesis testing, and the prior distributions for Bayesian techniques. The advent of Markov Chain Monte Carlo methods have unshackled Bayesian methods from the straightjacket of conjugate families, so we can perhaps focus here on the frequentist methods. The underlying population model for virtually all inferential techniques is the multivariate normal distribution, although occasionally one or two multivariate variants of the exponential and gamma distributions are used. Estimation is conducted almost exclusively using maximum likelihood, and estimators of all the common parameters are widely available. Likewise, the call for confidence regions rarely extends beyond those of standard parameters for which easily applied formulae are widely available.

The remaining area is hypothesis testing, and this is perhaps the most problematic area in multivariate inference. Following the usual principles, a null hypothesis typically specifies a single point in k-dimensional space for the unknown parameter, but now departures from the null hypothesis are not constrained as much as they were in univariate situations but can occur in many different ways (or directions in space). There are several different principles that can be employed in finding a test statistic in a multivariate situation, and unfortunately these principles often give different test statistics in any given situation. Not only are the test statistics different, but the inferences from these statistics may differ because they are all testing against different alternatives. Many standard computer packages will print out all the different statistics along with their significance levels for a given input of data, so the unwary user may find the apparently conflicting results rather confusing.

Perhaps, the two most commonly used general principles of test construction in the multivariate case are the likelihood ratio principle and the union-intersection principle. The former is based on a comparison of the maximised likelihood of the data assuming the null hypothesis to be true, with the unrestricted maximised likelihood of the data. It is designed to test the null hypothesis against *overall* departures, and the test statistic is generally referred to as *Wilks' lambda* in the computer output. By contrast, the union-intersection principle reduces the multivariate situation to a univariate one by taking a linear combination of the variables, and it finds the linear combination for which the null hypothesis is least appropriate. It thus tests the null hypothesis against the *maximum* departure, and the test statistic is usually referred to as *Roy's largest root* or *Roy's largest eigenvalue*. A set of data which disagrees with the null hypothesis moderately on all variables could end up as significant on the former test but not on the latter, while a set of data that disagrees markedly with the null hypothesis on just one variable and hardly at all on the others could end up as significant on the latter test but not on the former. Both tests (plus several others!) can be found in the output not only of specifically multivariate techniques such as canonical correlations, but also of multivariate extensions of univariate techniques, such as multivariate analysis of variance.

Missing values

A final comment should be made about missing values in a set of data. This problem can arise in most situations, but is perhaps most prevalent in multivariate data. For example, if the experimental units are plants or animals then some may die before all variates have been measured on them. In archaeology, artefacts or skulls may be damaged during excavation, or specimens may be incomplete when excavated, so not all variates can be measured on them. In surveys, a respondent may fail to answer some of the questions. Many multivariate techniques require complete records, so either (i) units or variates containing missing data have to be excluded from the analysis, or (ii) some estimate has to be made for each missing value.

Omitting complete units or variates is often wasteful of the data that have been collected, and can seriously reduce the sample size of the study. For example, if just 1% of the values are missing randomly and independently for each of 32 variates, omitting units that do not have all their values present will lose nearly 30% of the sample. So this action should only be considered if a particular unit or variate has an excessive number of missing values.

Estimating, or *imputing*, missing values is generally viewed as preferable. However, it does have some dangers. The simplest, if crudest, way of achieving it is to replace each missing value by the mean of all values present on the corresponding variate (for a given group if the data are grouped). The problems here are that this will reduce the variability of those variates that have many missing values, and it will falsely inflate the similarity of individuals that have missing values on matching variates. One possible form of remedial action is to add to each imputed value a random observation from a normal distribution having mean zero and variance equal to the variance of the relevant variate. This will inflate the variability back to the required level, and will reduce spurious similarity between individuals. A much more sophisticated imputation scheme is to assume the data set come from a multivariate normal distribution, and to estimate the missing values by maximum likelihood using the E–M algorithm. This is an iterative process that initially estimates the missing value for each variate by the mean of the observations present, and then uses multiple regression to successively fit values to each missing value until a stable situation is reached. Once the missing values have been estimated, then the data can be analysed as if complete.

10 Special Topics

Introduction

This book has not been primarily concerned with the details of statistical techniques, but rather with the underlying philosophy behind them and with the inferential processes that make it possible for us to draw conclusions from their results. We have tried to use as broad a brush as possible, and to cover most of the general situations in which this statistical reasoning is applicable. However, there is one very fundamental assumption that must be satisfied in any given situation if this reasoning is to lead to valid conclusions, and this is that any measurements made on different individuals/units must be independent of each other. In other words, the value(s) observed on one unit must not exert any influence on the value(s) observed on any other unit. The process of random sampling was expressly designed to ensure such independence of units, and all the inferential theory of Chapters 4 and 5 rests on it. Of course, different variables measured on the *same* units will always be associated, and the methods discussed in Chapters 6–9 can cope with such associations. Nevertheless, the assumption of mutual independence of the units must still be satisfied.

However, various situations arise where statistical analysis is required but this assumption is patently violated. An obvious case is when observations are repeatedly taken on the same individual as, for example, in a medical context where a patient's temperature and blood pressure are monitored at regular intervals over a period of time. Since the values are being observed for the same patient, successive observations are bound to be correlated, perhaps highly so. On a larger scale, government statisticians are much concerned with regular collection of data on such indicators as unemployment rates, house prices or money supply. Again, since it is the same quantity being monitored, the value at any one time of measurement is bound to influence the value at the next time of measurement. Moreover, 'time' can easily be replaced in this context by 'location'. The incidence rates of a cattle infection, say, for neighbouring localities are inevitably going to be more alike than those for distant localities. In all these situations, it would be indefensible to build a statistical analysis on the assumption of independence; the associations between units must clearly be taken into account in some way, and the previous reasoning processes must be modified.

There is another group of situations for which appropriate methods of analysis would not fit directly into the pattern of the previous chapters, but this time more because of the nature of the observations taken rather than because of a violation of assumptions. Much of the statistical inference that is conducted in everyday situations is concerned with inference about either the averages of populations or with the spread of values within them, because these are the primary descriptors of the populations and hence the aspects that most researchers are interested in. Thus many of the basic statistical methods are focused on either averages or spreads of values, where the individual values are taken to be independent observations from a common population. However, interest in some studies is more concerned with *extremes*. The environmentalist is interested in the highest level a river might reach, to assess the likelihood of floods; the engineer requires a bridge to withstand the strain of the anticipated loads on it, so is interested in the minimum breaking strengths of the constituent materials; the medical researcher seeking a cure for a serious illness will study carefully the minimum survival rates of patients on different treatment regimes, and so on. Such situations do not fall into the standard patterns of analysis, so again need more specialised treatment.

Of course, the relevant methodology to handle all situations such as those above has been extensively developed, and whole books are devoted to each topic mentioned. It is neither our aim, nor indeed would it be feasible, to cover the relevant methodology or to even give any of the details. Rather, in this final chapter, the aim is simply to sketch out the main features of a range of such specialised topics, to point out how they might depart from the standard templates of analysis discussed earlier, and to signal to the novice the need for expert advice should any of these situations be encountered. We specifically highlight a couple of the more common topics within each of the areas of correlated observations and analysis of extremes respectively.

Repeated observations

Since measurement errors are generally considerably lower in magnitude than the variability between individuals, the most accurate way of studying the effects of treatments over time is to measure each individual in the experiment repeatedly at different times and then to study the *profiles* of these individuals over time. An obvious example is in the medical treatment of a chronic disease where the effects of treatments may only become noticeable over a period of time, so researchers comparing several treatments will allocate each patient to one of the treatments, measure each patient repeatedly at specified time intervals, and then compare the sets of profiles for each treatment in order to establish whether there are differential treatment effects. However, this basic scenario is also surprisingly prevalent in many other different areas. Behavioural psychologists, for example, wishing to study the process of learning in animals will track each individual animal over time and measure some appropriate response (such as the time taken by a rat to negotiate a maze and reach the food reward) at each time point. Educationalists wishing to compare the effectiveness of different teaching methods will measure the proportions of test questions answered correctly by each subject under each teaching regime as time progresses. Animal health researchers developing a treatment for a parasite that affects the growth of cattle over the grazing season will need to monitor the weight of each cow at various times throughout this season.

There need to be at least three repeated observations made on each individual for the data to come within this heading, because if only two observations have been taken then most questions of interest can be answered by subtracting one value from the other and then analysing the resulting uni-variate differences between the two using standard methods. Moreover, typical scenarios do not have very many repeated measures—somewhere between about six and a dozen would be common. Also, typically there will be several treatment groups to compare, or even a more structured arrangement of the individuals, so a linear model analysis of variance format is strongly suggested. Within such a format, each individual could be viewed as forming a 'block', with the different times at which the measurements have been taken constituting levels of a 'time' factor. However, the observations within each of these 'blocks' are no longer independent, so what can be done?

The earliest approach was essentially to treat such a situation as a 'split-plot' arrangement (Chapter 6), with individuals as the 'whole plots' and the repeated measurements within each individual as the 'subplots'. Such an arrangement formally takes account of the correlations existing between the subplot values, and provides a valid mechanism for testing hypotheses and comparing groups. However, there are some drawbacks in adopting this arrangement. In a true split-plot design, the treatment levels are assigned at random to the subplots, but with time as the factor at the subplot level there can be no question of randomly permuting the time points!

More seriously, the linear model assumptions underlying the split-plot design imply that, although the repeated measurements are not treated as being independent, the correlation between all pairs of them are assumed to be equal. This is known as the *compound symmetry* condition; while it may be perfectly acceptable if there are only a few (e.g. three) repeated measurements on each individual, it becomes less realistic as the number of repeated measurements (and hence the elapsed time between different pairs of observations) increases. Nonetheless, it is employed in quite a variety of circumstances, and is now more familiarly known as *repeated measures ANOVA*. In fact, the compound symmetry condition is a special case of the more general *sphericity* condition. This latter condition has a rather technical definition, but there is a fairly simple test that can be applied to check whether any given set of repeated measurements satisfies it. If the data fails the test, then an adjustment can be made to the F test in the repeated measures ANOVA in order to compensate for the violation of its underlying assumptions. Many standard statistical computer packages include an option for performing the test of sphericity and conducting a repeated measures ANOVA.

Although the above process does provide a mechanism for analysing repeated measures, many researchers feel uneasy about the fairly restrictive assumptions that have to be made. An alternative approach is to treat the battery of repeated measures on each individual as a vector of multivariate observations, and to analyse the data using multivariate analysis of variance (MANOVA). In this case, the worry about the lack of randomisation of the time points disappears, because each point is now being treated as a fixed variable. Moreover, there are no longer any conditions imposed on the correlations between pairs of repeated measures—the data determine these correlations. Consequently, many researchers prefer this line of approach, on the grounds that it is much more generally applicable.

The objection levelled at the MANOVA approach, however, is that it is perhaps *too* general: no conditions are imposed, and the data effectively determine all the characteristics of the analysis. The problem with this is that there are thus very many parameters that have to be estimated (all inter-variable correlations as well as all the means and variances of the previous approach), and if the sample sizes are not very big then the power of the tests will not be large. Sharp conclusions are therefore less likely to be reached. In general terms, if data satisfy models that have relatively fewer parameters to be estimated, then the consequent hypothesis tests are more powerful. So in recent years many new methods for analysing repeated measurements have appeared, each having as its basis a relatively parsimonious model (i.e. one with few parameters) for the inter-measurement correlations. These models all aim to strike a compromise between the restrictive assumption of the repeated measures ANOVA and the lack of assumptions of the MANOVA. However, the methods deriving from these models are technically more complex than the two standard approaches, so we do not pursue them any further here. The reader is directed for further details to the references at the end of the book.

Time series

A time series also comprises a set of observations made on the same quantity at different points of time, but can be distinguished from repeated measures data in a number of ways. Instead of the measurements being made at an individual level, they are made at an aggregate or collective level: for example, the number of unemployed males between the ages of

18 and 65 at a given point in time, the number of passenger miles flown by an aircraft in a given month, the profit made by a large company at the end of the financial year, and so on. Typically, observations are made at regular intervals: daily or weekly for quickly changing processes (such as the monitoring of carbon by-products in the industrial production of a chemical), monthly or quarterly for more slowly evolving processes (such as, perhaps, sales figures or unemployment rates) and annually or every few years for a long-term perspective (usually for such things as government census figures). Finally, whereas repeated measures data might contain about 10 or a dozen values for each individual, time series data are usually collected over a much longer time span: years, tens of years or even, for one or two historical data sets, hundreds of years.

A typical time series will contain a number of identifiable components. There is generally a long-term *trend*, reflecting the overall (average) level of the values as time progresses, which may be either a general increase, or a general decrease, or a combination of both. So, for example, the unemployment rate may gradually increase in a country until the government takes notice and introduces various measures in order to encourage job creation, whereupon the rate progressively declines again. Sometimes this rising and falling of level occurs repeatedly over quite a long period, in which case it might be more correctly viewed as a *cycle* (e.g. the trade cycle). Shorter-term cycles are common with many measurements, and constitute *seasonal* effects such as the rises in household expenditure on heating in autumn and winter as compared with spring and summer. Finally, there are the *residuals*, which represent what is left over after trend, cyclic and seasonal components have been accounted for.

Simple *description* of a time series essentially consists of estimating each of these components at the various time points at which the measurements have been made. The estimation of the components depends on whether they are assumed to combine additively or multiplicatively. An additive model is indicated if the residual fluctuations are of comparable size irrespective of the value of the quantity being measured, while a multiplicative model is more appropriate when the fluctuations are proportional to the measured values. The trend value is generally estimated at each time point by a moving average (the average of a given number of contiguous observations centred at that time point), and the data values are then *detrended* either by subtracting the trend values from them (if the model is additive) or by dividing them by the trend values (if the model is multiplicative). Cyclic and seasonal components are then estimated by averaging the appropriate detrended values and normalising the results so that their average is zero (if the model is additive) or one (if it is multiplicative). Finally, removing these components from the detrended values by the same process as before leaves the residuals. The commonly used term 'seasonally adjusted' data refers to the application of this process to the original values (*not* to the detrended ones).

However, if any form of prediction of future values is contemplated, then it is necessary to predict the residuals at the relevant time points, and the feature of a time series that has to be taken into account is that the residuals are correlated. This is because any observation on the time series at one point will inevitably influence the value at the next point at which the measurement is taken—a sort of 'carry over' effect. Thus, in order to achieve accuracy of prediction, a realistic model must be formulated for the residuals and efficient estimation methods must be developed for the parameters of this model. This is what introduces the sophistication into the area of time series analysis, and the reason why whole books have been devoted to the topic.

In brief, the residuals are commonly modelled through either an *autoregressive (AR)*, or a *moving average (MA)*, or a mixed *autoregressive–moving average (ARMA)* model. An AR model assumes that an observation (i.e. residual) at time t is a linear combination of the values at a number of immediately preceding time points plus a random departure term; if the number of preceding values specified by the model is p then it is said to be an AR(p) model. An MA model assumes that an observation at time t is a linear combination of a number of independent random departure terms; if the number of such terms specified by the model is q then it is said to be an MA(q) model. As is then obviously implied, an ARMA(p, q) model is one where the observation at time t is assumed to be a linear combination of the values at the p immediately preceding time points plus a linear combination of q independent random departure terms. The process of reinstating trend, cyclic and/or seasonal components to predicted residuals is termed *integrating* the model, so the whole model-fitting exercise is generally referred to as ARIMA modelling of a time series. As the reader may imagine, there are many technical aspects relating to the choice of numbers p, q of ARIMA components, to the estimation of parameters of the model, and then to the estimation of likely errors of prediction.

It was suggested in the introduction that 'time' can be easily replaced by 'location' because measurements made in space, like those made in time, are almost certain to be correlated. This is true, but considerably undervalues the extra complications introduced when dealing with spatial rather than temporal data. On the (not unreasonable) assumption that all spatial data can be taken as occurring in two dimensions, it is immediately evident that correlations between neighbouring observations in space may depend critically on the *direction* between them. Thus the correlation between such neighbours in space is two-dimensional, whereas that between contiguous observations in time is one-dimensional—depending just on the time interval between the observations. One can simplify matters by imposing a further constraint on spatial data and requiring the correlation between two neighbouring observations to depend only on the spatial *distance* and not direction between them. This is the so-called assumption of *isotropy*, but not all data sets are amenable to such a constraint. Even if the correlation structure can be made more similar to the temporal one by this device, there are many more different forms of spatial data than there are of temporal data and this raises the need for a greater variety of sophisticated techniques for analysing them. For example, in epidemiological studies we may note the positions (i.e. spatial coordinates) of incidences of a particular disease, and interest focuses on determining whether such a *point pattern* can be taken as randomly distributed over the study area or whether it shows signs of clustering. A question much in the media at one time was whether leukaemia incidences were clustered round nuclear power stations, the alternative hypothesis being that the latter were simply present in the midst of a random distribution of cases. In agricultural experiments we may wish to allow for the effect of neighbouring plots on the treatment response in each plot. Here we need to use the correlations between neighbours to adjust the results of standard ANOVA or regression calculations. In environmental studies, we may only have the resources to measure the amounts of various trace elements present just at a few sites in a study area, and then wish to predict their amounts at other sites where there was not enough resource to measure them. In such a case, we would need to use the spatial correlations in modelling the amounts of the elements across the whole study area.

There are thus rather more problems to tackle in the spatial area than perhaps in the temporal one. A great deal of attention is given to the measurement and modelling of the spatial correlation in any practical application. Having done this successfully, the question then arises as to how this correlation should be employed in the analysis of the data, and various stochastic models have been proposed to deal with the different possible scenarios. Needless to say, such models involve some very sophisticated mathematics. Finally, as if dealing with spatial and temporal data separately is not enough, there are some problems (such as the study of the spread of epidemics) that require simultaneous handling of both of these factors. It should be clear from even this very brief mention that the field is a very rich one for statistical work, and that there are many problems that standard methodology will not cope with.

Analysis of extremes

In some disciplines, interest focuses particularly on predicting the occurrence of 'unusual' events, which leads to the recording of *extreme values* over given periods of time. The most common type of extreme is the *maximum*, examples being the maximum annual rainfall at a location, the maximum annual sea level at a location and the maximum annual height of a river at a location. Questions that might be asked here include: Is there a trend? What is the chance of a flood next year? Has some meteorological phenomenon had an effect? However, in some cases the interest may focus on the *minimum*, for example, the breaking strengths of glass fibres to determine maximum loads that a machine can withstand safely, or the fastest annual times of athletics events to study their rates of improvement.

In previous chapters we have seen that a probability model is essential for analysing data, and the emphasis of statistical inference has fallen mostly on either the probability function $P(X = r)$ if the measurement X is discrete, or on the probability density function $f(x)$ if it is continuous. In extreme value analysis, by contrast, the emphasis falls on the *distribution function* $G(u)$, which is the probability that X is less than or equal to u. If X is discrete then $G(u)$ is just the sum of all probabilities $P(X = r)$ for r less than or equal to u, while if X is continuous then it is the area under the probability density curve to the left of u. The reason why the focus is on the distribution function is because the simplest specification of the distribution of either the maximum or the minimum of a set of independent values from a given model is given by relating the distribution function of the maximum or minimum to the distribution function of the underlying model. This is because the probability that the maximum of a set of n observations is less than u is equal to the probability that *all* the observations are less than u, and if the n observations are independent with each having distribution function $G(u)$, then the probability is $\{G(u)\}^n$. Thus if $H(u)$ is the distribution function of the maximum, then $H(u) = \{G(u)\}^n$. Similar reasoning applies when seeking the probability that the minimum of a set of n observations is greater than u, which is equal to the probability that *all* the observations are greater than u. If $H(u)$ now denotes the distribution function of the minimum, then it readily follows that $H(u) = 1 - \{1 - G(u)\}^n$.

To tackle the analysis of extreme values, we therefore need to formulate a realistic model for them and this in turn will provide a model for $H(u)$. The old-fashioned approach (where by 'old-fashioned' is meant around the middle of the twentieth century!) was to choose one of the Gumbel family, the Fréchet family or the Weibull family of distributions as a model for the extreme value distribution, but more recently the three families have been subsumed into a single model, known as the *generalised extreme value (GEV) distribution*. The distribution function $H(u)$ for this model is given by the rather complicated-looking formula

$$H(u) = \exp\left\{-\left[1 + \xi\left(\frac{u - \mu}{\sigma}\right)\right]^{-1/\xi}\right\},$$

which has three parameters μ, σ, ξ representing location, dispersion and scale, respectively. Despite the rather daunting appearance of this formula, inference for a set of data consisting of a sequence of observed extreme values is well developed and software programs exist for either classical or Bayesian approaches.

On the classical front the best approach appears to be via maximum likelihood, where estimates of the parameters are readily obtained by numerical maximisation of the likelihood. Suppose for definiteness we have collected the maximum height of a river each year for n years. We use these values to estimate the three parameters μ, σ, ξ, substitute the estimates into the formula above to obtain the fitted GEV distribution function $H(u)$, and from this function calculate directly the values $z^{(p)}$ for different p, where p is the probability that $z^{(p)}$ is exceeded in any given year. These are the so-called *return levels* associated with the *return periods* $1/p$, and they form the most common summary measures. To obtain the equivalent summaries from a Bayesian approach, we need to formulate prior distributions for all three parameters, combine them with the likelihood to give the posterior distribution, use Markov Chain Monte Carlo methods on this posterior distribution to derive the predictive distribution of $H(u)$ and hence arrive at the Bayesian equivalent of $z^{(p)}$.

As with our other categories in this chapter, many extensions and sophisticated variations on the above ideas have been developed, but we cannot go into them here. The important message is that if extreme values constitute the observed data, the reader should be aware that standard methods based on the normal distribution are most certainly *not* appropriate.

Survival data

Survival data comes in various guises, but the essential features are that the measurement is *time*, from a well-defined starting point (the start of the observation) up to another well-defined point event. The latter is often called the 'failure', but it can equally be thought of as simply the end of the observation period, or in various equivalent terms. The most obvious example is the set of survival times until death of patients in a clinical trial, but a less dramatic example in medical statistics would be the set of times until cure of a particular treatment. An example in industrial reliability would be the times to failure of machine components. Editors of scientific journals often provide annual reports in which they discuss the times taken from receipt of articles to the completion of the reviewing process by chosen referees, and another example in a similar vein would be provided by a library survey of borrowing times, from the issue of a book to its return by the borrower. Finally, the duration of strikes would be an example in Economics.

Interest sometimes focuses on a single group or population, in which case it is a description of that group or perhaps a test of a simple hypothesis that is required. For example, the library might simply want to have a confidence interval for the mean time that a randomly borrowed book is off the shelves, or the medical researcher might want to test the hypothesis that the mean time to cure for a particular treatment is greater than three months. On the other hand, a comparison between two or more groups is often of prime interest. For example, the author of a scientific paper might wish to submit it to the journal that has the quickest average review time, so wishes to compare the averages of a selection of journals, or the industrialist wants to compare the times to failure of several different makes of components. Moreover, there are often many potential *covariates* that might influence the survival times, and the researcher may be interested in determining which of them has the greatest effect.

The reader may be forgiven for wondering at this point what the problem is; can we not just use the reasoning of Chapters 4 and 5 for inferences on either a single population or a set of populations, and the methods of Chapters 6 and 7 for a study of the effects of covariates? The answer of course is yes, we can, but only if *we have full information on all measurements*—and in most studies involving survival data we do not have such full information. The reason we do not is because most studies or data gathering sessions have to be completed in a set time, and some (often many) observations are not completed by this time. In the clinical trial, some patients may still be on a treatment when the study comes to an end; if such a patient has been taking the treatment for time *t* at this stage, then all we can say is that his or her survival time is *greater than t*. If a book is still out on loan when the library survey comes to an end, then that book's borrowing time is similarly greater than *t*; and likewise for durations of strikes that are still in progress at the end of the study, or scientific articles that are still with the reviewers when the editor writes the report, and so on. Such observations are said to be *censored*; instead of having a precise value *t* for them, all we can say is that it is some value greater than *t*. (Strictly speaking they are *right-censored*, because the actual value is greater than the cut-off point. Some situations arise where observations are *left-censored* if values *below* some threshold cannot be observed. For example, a weighing machine might not register very small weights, less than say, *a*. Then a weighing of 'zero' can actually be anywhere between 0 and *a*. However, we do not include these cases in the present discussion). The point about survival data is that censoring needs to be taken into account, and standard methodology has to be modified in order for this to be done.

Some early work focused on nonparametric methods. Various estimators of mean survival time were developed, the most popular probably being the Kaplan–Meier estimator, which continues to be used in many studies to the present day. However, for wider inferences and more power, the emphasis has increasingly been on parametric methods. A parametric approach requires a choice to be made first for the population model of the survival times, and various probability models are available. The exponential, gamma and Weibull models were mentioned in Chapter 3, and these are popular choices, while other possibilities include the log-normal, the compound exponential and the log-logistic models. Choice between them can often be guided by their *hazard* functions, which are functions giving the instantaneous probabilities of failure at any time point. The physical situation being studied will often suggest what the behaviour of the hazard function should be like, and this will suggest the most appropriate model(s) to use. For example, if the lifetimes of projector bulbs are being studied, then the hazard function is large for the first few seconds of use (surge of power at the start may blow the bulb), then it drops rapidly to a relatively low and approximately constant rate for a long time (low chance of failure while operating normally) but then rises again as the bulb ages (increasing chance of failure due to 'old-age').

Once we have chosen a probability model, maximum likelihood estimation of the parameters, and inferences following from it, are fairly straightforward. The contribution to the joint probability of the data of those observations that have precisely measured survival times are given simply by the probability density functions at those survival times. The contributions of the censored observations are derived from the distribution function. If $G(u)$ is the distribution function of the probability model, then $G(u)$ is the probability that the survival time of an individual is less than or equal to u. Thus if an observation has been censored at time t then its contribution to the likelihood must be $1 - G(t)$, namely the probability that the survival time is greater than t. Standard maximum likelihood methods are then applicable to the resulting likelihood.

Dependence of survival times on explanatory variables has increasingly become the focus of attention, and many different models have been proposed and studied. By far the most popular is the proportional hazards model, with various adjustments for special cases. However, as with our other topics in this chapter, we stop short of discussing the technicalities of these models.

Conclusion

The four topics briefly considered above are just a few of those that seem, on the face of it, to be amenable to standard statistical methodology—but the more that one digs beneath the surface the more it becomes evident that special methods have to be devised to tackle them. This really is the story of the development of statistical methods in a nutshell. In any situation, simple models are first proposed and correspondingly simple analyses are derived from them. Critical examination of the models and results inevitably bring shortcomings to light, and the models then have to be re-examined and modified. This in turn will often lead to modified analyses. But however far the methodology is developed, it is never enough—situations always arise which require further development. That is the enduring fascination of Statistics.

Sources and Further Reading

Much of the content of this book has been informed by the many textbooks I have at least dipped into over the years, and some of these I recommend below for further reading. However, I first need to mention those books that I specifically consulted and that were particularly useful during the actual process of writing. As a source of reference for technical matters, the two volumes of Kendall's Advanced Theory of Statistics devoted to classical statistics (Stuart and Ord, 1987, 1991) and the volume devoted to Bayesian statistics (O'Hagan, 1994) were invaluable. I also made much use of Stigler (1986) for the various historical comments, and of Marriott (1990) for accuracy of statistical terminology.

Turning to books for further reading, the recommendations are arranged roughly in chapter order and are kept, as far as possible, to texts that avoid very technical or very mathematical developments.

There is now a glut of introductory textbooks on the market, all of which cover to varying levels of technical detail and comprehensiveness the material of Chapters 1, 2, 3, 4 and 6, but it is not easy to guess the level of mathematical difficulty in a randomly chosen title. As a general guide, the reader looking for a non-technical and simple exposition would probably be advised to find an introductory text aimed at a specific subject area, with titles such as 'Introductory Statistics for Psychologists', 'Statistics for Biologists', and so on. Two nice examples of this type are the ones by Anderson (1989) and Diamond and Jefferies (2001), the older one by Rowntree (1981) can still be recommended, and a very readable account focusing on semi-recreational aspects of probability is the book by Everitt (1999) quoted in Chapter 1. If the reader is looking for greater mathematical and computational detail, then Daly *et al.* (1995) gives a wide-ranging coverage arranged roughly like the present book, a similarly comprehensive account is given by Clarke and Cooke (2004), but a book such as that by Hogg and Tanis (2006) will appeal only to the mathematically competent.

Likewise, books devoted to statistical inference, whether classical or Bayesian, generally require a considerable degree of mathematical sophistication. The adventurous might dip into Lee (2004) or Migon and Gamerman (1999), but should not expect an easy ride!

Many of the introductory books listed above also cover the basics of regression modelling and analysis of variance as developed in Chapter 6, but few if any proceed to the more advanced features of regression or to the generalised linear models of Chapter 7. Krzanowski (1998) straddles the two chapters to some extent, while Dobson (2002) focuses more on Chapter 7. Similarly, much of the detail of material in Chapters 8 and 9 is mathematically quite sophisticated, but there are various general texts that explain the underlying reasoning while focusing principally on analysis of multivariate data using existing computer software. Bartholomew *et al.* (2002) do a good job of discussing the techniques and their uses while avoiding mathematics almost completely, Manly (2005) gives concise and relatively simple expositions of the main techniques, Afifi *et al.* (2004) concentrate almost exclusively on the computational interpretations, while Krzanowski (2000) has more emphasis on traditional mathematics. Everitt (1984), Everitt *et al.* (2001) and Hand (1997) are useful texts devoted to the specific areas of latent variable models and classification.

Each of the various special topics contained in Chapter 10 has given rise to numerous books, but many of them are very technical and detailed so here we content ourselves with just one book on each area covered. Crowder and Hand (1990) give a readable account of repeated measures analysis; Chatfield (2004) provides a good introductory overview of time series methodology; Bailey and Gatrell (1995) describe a practical approach to spatial analysis; Coles (2001) is an authoritative (but technical) text on extreme value theory; while Collett (2003) provides a good introduction to survival analysis. Each of these books lists many other references that interested readers can follow up.

References

A. Afifi, V.A. Clark and S. May (2004) *Computer-Aided Multivariate Analysis*, 4th Edition. Chapman and Hall/CRC, Boca Raton, FL.

A.J.B. Anderson (1989) *Interpreting Data: A First Course in Statistics*. Chapman and Hall, London.

T.C. Bailey and A.C. Gatrell (1995) *Interactive Spatial Data Analysis*. Longman Scientific and Technical, Harlow, Essex.

V. Barnett (1974) *Elements of Sampling Theory*. The English Universities Press Ltd., London.

D.J. Bartholomew, F. Steele, I. Moustaki and J.I. Galbraith (2002) *The Analysis and Interpretation of Multivariate Data for Social Scientists*. Chapman and Hall/CRC, Boca Raton, FL.

J.M. Bland and D.G. Altman (1994) 'Regression towards the mean', *British Medical Journal*, Vol. 308, p. 1499.

C. Chatfield (2004) *The Analysis of Time Series, an Introduction*, 6th Edition. Chapman and Hall/CRC, Boca Raton, FL.

G.M. Clarke and D. Cooke (2004) *A Basic Course in Statistics*, 5th Edition. Arnold, London.

S. Coles (2001) *An Introduction to Statistical Modeling of Extreme Values*. Springer, London.

D. Collett (2003) *Modelling Survival Data in Medical Research*, 2nd Edition. Chapman and Hall/CRC, Boca Raton, FL.

M.J. Crowder and D.J. Hand (1990) *Analysis of Repeated Measures*. Chapman and Hall, London.

F. Daly, D.J. Hand, M.C. Jones, A.D. Lunn and K.J. McConway (1995) *Elements of Statistics*. The Open University and Addison-Wesley, Wokingham.

I. Diamond and J. Jefferies (2001) *Beginning Statistics: An Introduction for Social Scientists*. Sage Publications, London.

A.J. Dobson (2002) *An Introduction to Generalized Linear Models*, 2nd Edition. Chapman and Hall/CRC, Boca Raton, FL

B.S. Everitt (1984) *An Introduction to Latent Variable Models*. Chapman and Hall, Boca Raton, FL.

B.S. Everitt (1999) *Chance Rules: An Informal Guide to Probability, Risk and Statistics*. Springer-Verlag, New York.

B.S. Everitt, S. Landau and M. Leese (2001) *Cluster Analysis*. Arnold, London.

R.A. Fisher (1925) *Statistical Methods for Research Workers*. Oliver and Boyd, Edinburgh.

R.A. Fisher and F. Yates (1938) *Statistical Tables for Biological, Agricultural, and Medical Research*. Oliver and Boyd, Edinburgh.

F. Galton (1886) 'Regression toward mediocrity in hereditary stature', *Journal of the Anthropological Institute*, Vol. 15, pp. 246–263.

D.J. Hand (1997) *Construction and Assessment of Classification Rules*. John Wiley & Sons Ltd., Chichester.

R.V. Hogg and E.A. Tanis (2006) *Probability and Statistical Inference*, 7th Edition. Pearson Education, New Jersey.

H. Joyce (2002) ' Beyond reasonable doubt', *Math Plus*, Vol. 21.

W.J. Krzanowski (1998) *An Introduction to Statistical Modelling*. Arnold, London.

W.J. Krzanowski (2000) *Principles of Multivariate Analysis: A User's Perspective*, Revised Edition. Oxford University Press.

P.M. Lee (2004) *Bayesian Statistics: An Introduction*, 3rd Edition. Arnold, London.

B.F.J. Manly (2005) *Multivariate Statistical Methods, a Primer*, 3rd Edition. Chapman and Hall/CRC, Boca Raton, FL.

F.H.C. Marriott (1990) *A Dictionary of Statistical Terms*, 5th Edition. Longman Scientific and Technical, Harlow, Essex.

H.S. Migon and D. Gamerman (1999) *Statistical Inference, An Integrated Approach*. Arnold, London.

J. Nelder and R. Wedderburn (1972) 'Generalized linear models', *Journal of the Royal Statistical Society Series A*, Vol. 135, pp. 370–384.

A. O'Hagan (1994) *Kendall's Advanced Theory of Statistics: Volume 2B, Bayesian Inference*. Edward Arnold, London.

K. Pearson (1914) *Tables for Statisticians and Biometricians*. Cambridge University Press.

D.F. Percy (2005) 'The power of Bayes', *Mathematics Today*, August, pp. 122–125.

D. Rowntree (1981) *Statistics Without Tears. An Introduction for Non-mathematicians*. Penguin Books, London.

A. Stuart and J.K. Ord (1987) *Kendall's Advanced Theory of Statistics: Volume 1, Distribution Theory*, 5th Edition. Charles Griffin & Company Limited, London.

A. Stuart and J.K. Ord (1991) *Kendall's Advanced Theory of Statistics: Volume 2, Classical Inference and Relationship*, 5th Edition. Edward Arnold, London.

S.M. Stigler (1986) *The History of Statistics: The Measurement of Uncertainty before 1900*. Harvard University Press, Cambridge, MA.

W. Yule, M. Berger, S. Butler, V. Newman and J. Tizard (1969) 'The WPPSI: an empirical evaluation with a British sample', *British Journal of Educational Psychology*, Vol. 39, pp. 1–13.

Index

Printed in the United States
By Bookmasters